WHAT OTHERS ARE SAYING

The written words are so powerful, insightful, and mind-altering. This is a fantastic book—creative and inspirational."

—Dr. Janet Holdcraft,
assistant superintendent
of Glassboro Public Schools

I have had the privilege of knowing Wellington Watts for over 30 years. He is one of the most consistent followers of Christ that I have known. He is deeply spiritual and has a love for God's word and is committed to the cause of Christ. His exceptional ability in explaining God's word is evident in this book. I know you will be blessed and encouraged in reading it.

—Rev. D. Donald Brasco,
pastor of Calvary Hill Church for thirty-six years

Wellington Watts is a modern day Moses—a leader who is not afraid to stand for truth and righteousness in times of adversity, yet one who is meek in all of his ways.

—Kathleen Berkheiser,
school board member and
former assistant prosecutor

What a wonderful project. If social networks are so in vogue, why not a social network that helps people with prayer. Get it done! Good luck with your book. So glad you are working, thinking, publishing, living, and doing well.

—Love, Cheryl Kramer Hoey,
a classmate (class of 1960) at Milton High School,
Milton, Delaware

THE IMPACT OF TECHNOLOGY
3D IS COMING

THE IMPACT OF TECHNOLOGY 3D IS COMING

MAKING EDUCATION COME ALIVE

WELLINGTON E. WATTS II

TATE PUBLISHING
AND ENTERPRISES, LLC

Published by Tate Publishing & Enterprises, LLC
127 E. Trade Center Terrace | Mustang, Oklahoma 73064 USA
1.888.361.9473 | www.tatepublishing.com

Tate Publishing is committed to excellence in the publishing industry. The company reflects the philosophy established by the founders, based on Psalm 68:11,
"The Lord gave the word and great was the company of those who published it."

Cover design by Nikolai Purpura
Interior design by Caypeeline Casas

Published in the United States of America

ISBN: 978-1-63367-447-9
1. Technology & Engineering / General
2. Education / Teaching Methods & Materials / Science & Technology
14.11.10

I am dedicating this book to God, my source of life, for keeping me supernaturally alive for His purpose.

Thank you, God, for your revelations and Your divine interventions in my life, for You do all things well!

I am also dedicating this book to my wife, Nancy, and son, Wellington the Third, who have been a source of encouragement for me for many years.

A special welcome to the "nones"—the religiously unaffiliated—and to the millennials.

May the information in this book give you another perspective for living.

ACKNOWLEDGMENTS

A Special Tribute to My Friend Richard Jones

A week before Christmas, I received a special gift from Richard's wife, Janet: a Louisville Slugger bat with George "Babe" Ruth inscribed near the end. What makes this bat so special is that it is shaped like a cane with a rubber tip. It was like the same cane Richard used the last time I saw him at a Phillies game.

Richard was born with cerebral palsy and never could walk like most people. He shuffled his way through life. His father had died before I met him, and he was raised by a single parent. But I never heard him complain about his life.

I was ten when we moved to Milton, Delaware, when I first met Richard. That means we have been friends for sixty years. We would ride our bikes to Milton Camp in the summer. I spent Saturday mornings at his house, watching football games, listening to Lindsey Nelson, and drinking Frosty Root Beer. Since we didn't have a TV at the time, we would go to Richard's house to watch *I Love Lucy* and *The $64,000 Question*, hosted by Hal March.

Richard came to our wedding, and I was the best man in his wedding. We would meet annually at the Vet and then at Lincoln Financial Field for a Phillies/Dodgers game, followed by a stop at a diner. Though handicapped, the state of Delaware gave him a career monitoring traffic flow at designated intersections.

And then Richard couldn't drive anymore. Richard couldn't work anymore. The annual meetings at the Phillies games stopped.

Richard, with Alzheimer's disease, is now in a nursing home. The last time I called him, he didn't know who I was. But I know who he is, and the friendship we had will never end.

The bat cane hangs on the door of our entertainment center, and each time I see, it I remember my friend, and a little smile emerges when I think of those times we had together, when there was laughter, when we had fun.

A special thanks to God for His great love and mercy to me and for the miracles He has provided me making this entire endeavor—this book, *The Impact of Technology: Impacting Education and Impacting the Church*—possible.

All my successes belong to Him. All my failures belong to me. My choices have been the difference between the two.

A special thanks to my parents, Rev. W. Earl Watts and my mother, Anna Watts, for the Christian heritage they gave me. Because of King Yeshua/Jesus Christ, we will meet again.

A special thanks to my wife for her love, for being a supportive wife over the years, for being the best mother, for being my very best friend—all this while enduring a lot of physical pain over the last twenty-five years. Without her, this would not have been possible.

Special thanks to my son, Wellington E. Watts III—the owner of Alexandria Colonial Tours, at HYPERLINK "http://alexcolonialtours.com/" who is the best son one could have. His Christian example and leadership in his community has demonstrated his Christian values and Christ-centered life.

Special thanks to Rev. D. Donald Brasco—pastor emeritus of Calvary Hill Church in Glassboro, New Jersey—and to his wife, Mary Jane Brasco. Pastor Brasco's ministry as a pastor and his support as a friend have been greatly appreciated.

Special thanks to my family members: Marilyn Watts, Skeete and Beverly Watts Nalley, Rev. Ronald Watts, and Candy Watts and family. Special thanks to my extended family members: Herb and Nellie Grose and family, Harry and Doris Worrell and family, and Harold and Margarethe Worrell and family for their love and support over the years.

Special thanks to Al and Barbara Frank for their friendship and support.

Special thanks to Joel Rosenberg for writing "Epicenter" and establishing a focal point for *The Impact of Technology: Impacting Education and Impacting the Church.*

Special thanks to Stephanie George Chambers for contributing her photograph.

Special thanks to Pastor Lon Solomon of McLean Bible Church for his inspiring Internet sermons that helped God give me additional insights in respect to a new role of the church.

Special thanks to the late Dorsey Marshall Jr., whose Face Book Success Minute touched many people, including me.

Special thanks to the late Rev. Tommy Holshouser

Eastern Pilgrim College made an exception. They allowed me to sing in the King's Ambassadors Quartet during the summer of 1960 before I was actually enrolled at the college. It was a big jump for me, just graduating from high school. Within three weeks, I was singing with the King's Ambassadors.

Tommy Holshouser was the public relations director for Eastern Pilgrim College. Our quartet traveled with Tommy Holhouser over many miles during the summers of 1960 and 1961, representing Eastern Pilgrim College. During those early college years for me, Tommy Holhouser was a special person to me and helped me with his encouragement and belief in me.

I think Tommy liked country music for he would sing some lines from songs he liked as we were riding those many miles. He had a way of teasing some of the guys and getting his message across—not me, of course. He knew there would be girls that we would encounter at our stops. He also knew that the guys already had girlfriends. His way of communicating his thoughts to the guys with "wandering eyes" was to begin singing, as we rode, "Crazy arms that long to hold somebody new." Everyone knew what he was doing, and there were the moments of sheepish grins. Caught!

He was a man of integrity, vision, and passion, which was demonstrated in his speaking. He loved his wife and daughter, Cathy, deeply. I am forever grateful for his contribution to my life.

CONTENTS

Introduction 15

Part I	Rethinking Education	17
	1 The Research	19
	2 Introducing ViziTech USA	27
Part II	Rethinking Business	91
	3 Millennials and Cell Phone Ad Campaigns	93
Part III	Rethinking Church	105
	4 The Research/The Millennials	107
Part IV	Rethinking Events	127
	Prologue to Understanding Events and Life	129
	5 The Perspective of Time	135
Part V	Rethinking Life	167
	6 How Shall We Then Live?	169
	About the Author	189
	Endnotes	205
	Appendix	207

INTRODUCTION

The Reality of Accelerating Change

My name is Wellington E. Watts II. I was educated at Eastern Pilgrim College with a BA degree. I took additional education courses at Glassboro State Teachers College, now Rowan University. I am a K-8 permanently certified New Jersey teacher. I taught in the public schools, a Christian school, and I was the principal of Ambassador Christian Academy. I am currently a technology researcher and analyst for Alexandria Colonial Tours.

New Jersey's Student-Growth Objectives and Technology

Student-growth objectives are long-term academic goals that teachers set for groups of students, and they must be:

1. Specific and measurable
2. Aligned to New Jersey's curriculum standards
3. Based on available prior student learning data
4. A measure of what a student has learned between two points in time
5. Ambitious and achievable
6. Achieve NJ

The problem with student growth objectives is that they are established in the context of an "educational stagnation" environment.

According to the US education secretary, students from other countries are surpassing students in the United States, according to the test results from the Program for International Student Assessment (PISA). US education secretary Arne Duncan characterized the US flat scores as a "picture of educational stagnation" (from a report by Daniel Arkin, staff writer, NBC News).

The utilization of technology creates the possibility of students exceeding student-growth objectives that have been established in a traditional educational format. That technology is the inclusion of 3D technology in the learning process.

PART I

Rethinking Education

1

THE RESEARCH

The 3D in Education White Paper
Written by Professor Dr Anne Bamford,
Director of the International Research Agency

Important Research Excerpts on 3D in Education

The research took place between October 2010 and May 2011 across seven countries4 in Europe. The study focused on pupils between the ages of 10-13 years learning science-related content. The research project involved 740 students, 47 teachers and 15 schools across France, Germany, Italy, Netherlands, Turkey, United Kingdom and Sweden. Equality of access is the law in Europe, so the schools included children from different backgrounds and with learning or behavioral challenges integrated into the general classes. The study involved: Private and public schools; single sex schools; city schools and rural schools; high and low academic achieving schools; technology-rich and technology-poor schools; large schools and small schools; primary, middle and secondary schools; and experienced and less experienced teachers. In each school there was a 'control' class and a 3D class. Both classes had the same instruction, but the 3D class also had the 3D resources.

Children and 3D

Children and young people own a lot of technological devices and use them regularly. As indicated by the recent pan European research2, 90.1% of pupils had a computer, 85.3% had at least one mobile phone and 74.6% owned handheld games.3 It also found that pupils are frequent users of online technology, with over 91% of pupils using the internet for at least one hour per day. In terms of their experience of 3D, 90% of pupils had seen a 3D movie, with most pupils having seen three or more 3D movies. The pupils were very knowledgeable about general innovations in 3D and were highly informed consumers of the 3D products currently available. The pupils possessed very positive attitudes towards 3D and were keen to have more 3D in their lives and in their learning. The teachers that were interviewed acknowledged the importance of good quality technology for the pupils of today, as they are "digital native" learners, as the following comments from teachers exemplify:

"The kids are into technology. We need something different in the classroom. It is more philosophical than just putting computer in the classroom. Technology is not just about learning the content. Technology will change the view of life. Children must have different points of view on life."—Teacher Comment

"The pupils wanted, and expected, very high quality animations."—Teacher comment

Why is 3D important?

Children find it hard to understand what is not visible. Visual learning improves the pupils' understanding of functionality and by seeing the whole of something; children are able to understand the parts.

The research results indicated that the pupils had a strong preference for visual and kinesthetic learning, with 85% of the pupils preferring seeing and doing, while only 15% of pupils preferred hearing.

"Teachers talk a lot and you just sort of tune out, but when you see things it is there and suddenly it all makes sense."—Pupil comment

Complex concepts become more easily digested when reduced to imagery. The research results suggested that the 3D animated models were able to represent information in the most economical manner to facilitate learning and comprehension, thus simplifying complex, abstract and impossibly large amounts of information into a coherent form. By rendering the world visually, the children were able to understand greater levels of complexity, as the animations allowed the pupils to see structures and to see how things worked. In particular, the 3D animations made it possible for pupils to move rapidly from the whole structure to various parts of the structure, including to the microscopic and cellular levels. This process of amplification and simplification seemed to be particularly effective as an aide to understanding.

The 3D content in the classroom appears to 'come out to' the pupils. The deepest 3D and the most animated content appeared to have the greatest effect on learning and retention. These highly vivid experiences make the learning very captivating to the senses. During class observations, 33% of the pupils reached out or used body mirroring with the 3D, particularly when objects appeared to come towards them and where there was heightened depth.

The impact of 3D on academic results

The results of the research indicate a marked positive effect of the use of 3D animations on learning, recall and performance in tests. Under experimental conditions, 86% of pupils improved from the pre-test to the post-test in the 3D classes, compared to only 52% who improved in the 2D classes. Within the individuals who improved, the rate of improvement was also much greater in the classes with the 3D. Individuals improved test scores by an average of 17% in the 3D classes, compared to only an 8% improvement in the 2D classes between pre-test and post-test. The marked improvement in test scores was also supported by qualitative data that showed that 100% of teachers agreed or strongly agreed that 3D animations in the classroom made the children understand things better, and 100% of teachers agreed or strongly agreed that the pupils discovered new things in 3D learning that they did not know before. The teachers commented that the pupils in the 3D groups had deeper understanding, increased attention span, more motivation and higher engagement. The findings from the teachers were also evident in the findings from the pupils, with a higher level of reported self-efficacy in the pupils within the 3D cohort compared to the 2D control groups.

"I think I will get better test results. It is easier for me to remember with 3D. Then I will do well."—Pupil comment

The pupils felt strongly (84% agreed or strongly agreed) that 3D had improved their learning. High levels of pupil satisfaction with 3D learning were also evident with an 83% approval rating.

The pupils in the 3D class were more likely to recall detail and sequence of processes in recall testing than the 2D group. Both pupils and teachers stated that 3D made learning more "real" and that these concrete, "real" examples aided understanding and improved results.

The 3D pupils were also more likely to perform better in open-ended and modeling tasks. During the research study, several tests were undertaken to test for regression. Teachers were asked to note the pupils' retention (memory) after one month, both in terms of qualitative and quantitative differences between the retention in the 3D-based learning and the non-3D-based cohorts. Open-ended tasks were given to determine the impact both on retention and on recall. The teachers noted changes in the manner in which the 3D and 2D pupils recalled the learning. For example:

• The 3D pupils were more likely to use gestures or body language when describing concepts •

The 3D learners had better ordering (sequence) of concepts •The knowledge of concepts was greater in the 3D cohorts (especially when a new concept had been introduced through 3D) •

The 3D cohort had enhanced skills in describing their learning including writing more, saying more and being more likely to use models to show learning.

"In this school we find that theoretical retention is a problem. As I see it, the 3D increases visual retention and this boosts learning."—School principal

"When the teacher shows a model if it is small you can't see it, but with 3D even if the teacher moves around or a big kid is in front of you the 3D will always move in front so you can always see things clearly."—Pupil comment

The impact of 3D on classroom interactions

The use of 3D in the classroom led to positive changes in pupils' behavior and communication patterns and improved classroom interaction. The "on task" conversations and questions in the classroom increased after 3D was seen in a lesson. The pupils in the 3D group were more inclined to ask complex questions. The pupils were highly motivated and keen to learn through a 3D approach. The teachers found that the use of the 3D technology led to a deepening of pupils' understanding, increased attention spans, more motivation and higher engagement.

"In class with 3D you have the 'Wow' effect. This helps with behavior. The pupils are too interested to be disruptive. They get involved and forget to be naughty! I would like to keep using it and use it for different topics."—Teacher comment

On average, 46% of pupils were attentive at five-minute interval tests during the non-3D part of teaching the lesson, compared to 92% of pupils being attentive at five-minute intervals during the 3D part of the lesson. Interestingly, when the 3D part of the lesson was over, attentiveness continued to rise and would remain high for the rest of the lesson. For example, 96% of pupils were attentive in the five minutes following the 3D.

It appears that the 3D experience and resulting questions continued to promote attentiveness. Boys and pupils with attention disorders showed the most positive change in attention levels and communication (including asking questions) between 2D and 3D.

"The class certainly pays more attention in 3D. They are more focused. That is important in this class–8 out of the 26 pupils in this class have attention problems, so I am thrilled with the impact of 3D. They sit up and are really alert."—Teacher comment

"3D in the lesson makes them concentrate more. They have to focus and concentrate."—Teacher Comment

"We are sure that the system should be in every school and be available for every teacher." Principal Comment 5

—Dr. Anne Bamford

Current Studies

Dear Wellington,

Thank you for the updates and glad to hear the expansion of your work. We are now doing an extended 3D trial in 13 schools in London. It is going well and already the schools are working together and collectively supporting the rollout…a really nice model.

The project is being evaluated so we can see the results. I will certainly keep you updated. I would be very happy to share this research in DC when things get going at your end.

All the very best,
Dr. Anne Bamford

2

INTRODUCING VIZITECH USA

The following information is provided by ViziTech USA.

> Vizitech USA is the nation's leading sales organization for enhanced 3D imaging in Medical, Education, Commercial and Military applications.

ViziTech's Team

Brigadier General Stewart Rodeheaver (Retired)

Stewart Rodeheaver, a retired brigadier general and president of ViziTech USA, began briefing the audience of media and well-wishers at the Atlanta Press Club's June luncheon that were convened to honor him and the men and women under his command—the 4,600 citizen soldiers of the Georgia Army National Guard's Forty-Eighth Brigade Combat Team.

"We went in with the mandate that we were going to treat everybody with dignity and respect," Rodeheaver says. "We were going with the premise that we are here to rebuild and secure, not secure and rebuild.

"So we didn't shoot first. We tried to work with people, because I really believe that you change the world on a person-to-person basis."

Rodeheaver always has been a consensus builder. He spent thirty years with Georgia Power, first in construction then in community and economic development, based in Macon, bringing cities, counties, and business interests together to work through development and planning issues.

As a general, he did those things in Iraq also. He sat with tribal leaders; preached the gospel of democracy, debate, and dialogue; and facilitated the construction of water treatment plants, schools, medical centers, and other infrastructure.

Service Disabled Veteran Owned Business

Founder Stewart Rodeheaver (brigadier general, retired) has over forty years' experience in the military, business, education, training, and technology fields. He and his team have made it their goal to help raise the bar for educational and training platforms in America.

> ViziTech USA—a breakout learning technology company involved in leading edge research and application of unique three-dimensional training environments. Our unique new 3D training technologies have been presented to and/or have been purchased by multiple interested government entities, including the Chairman of the Joint Chiefs of Staff, the United States Army Intelligence Center at Ft Huachuca AZ, the DNA and Forensics Training

> Department at the US Army Criminal Investigation Laboratory, the US State Department, the Georgia Department of Transportation, and others.
>
> —ViziTech Website, Internet

ViziTech's Chief Scientist

Dr. Carroll H. Lastinger, ScD, is ViziTech's chief scientist. He is a Georgia native and spent twelve years in the US Air Force, where the highlight of his military service was in direct support of the White House and president of the United States by transporting the presidential motorcade vehicles and Secret Service advanced teams for international US presidential engagements.

He achieved ground, flight, and simulator certified-instructor service throughout the majority of his career. He acquired pilot training and licensing through the US Air Force, with undergraduate degrees from the Air Force College undergraduate Jet Propulsion and Flight Engineering Aeronautics, and an FAA Airframe and Powerplant License.

As a flight simulator instructor for the US Air Force, he programmed and began to use computers and CAD extensively to explore new types of visualization of aircraft maneuvers, components, stress dynamics, and flight envelope limitation scenarios. This helped reduce airframe component fatigue factors and increase safety because of the clarity of instruction coupled with 2D/3D computer visualizations and animations.

While in the US Air Force, Dr. Lastinger realized the immense value of "visual based" training, and for individual purposes, he embarked upon personal research and development of computer-based spatial three-dimensional imagery using a PC-type computer instead of very expensive holography. His experience of virtual reality (VR) and its weak reviews solidified in his mind the need to develop and formalize a value-based spatial imagery process that would not cause physiological side effects, such as headaches and nausea.

After extensive private research, Dr. Lastinger formulated a combination of inseparable visual laws, perception cues, and environmental geometry into an immersive, seamed, three-dimensional algorithm that, when used, provides the viewer with an unrivaled ability to cognitively internalize a first-person-experience instead of a second-person perspective. This "leap" in visualization capability caused him to seek, and subsequently be awarded, a patent for this technology.

He further sought and was awarded the trademark 3D HoloProjection as the name of this technology. This body of research became his doctoral thesis with Emmanuel Baptist University, and he was confirmed in 2002 as a doctorate of science (ScD) degree in Visual Cognition.

Wanting to revitalize 3D HoloProjection, he partnered with seasoned and experienced business partners to further his life's work. Their vision of deploying this technology into multiple industries reignited his passion for providing a world-class paradigm shift in communications and media. Their business experience, coupled with our portfolio of technology products, positions our initiatives as the next success story rivaling Microsoft.

Dr. Carroll Lastinger currently holds the position of chief scientist for a corporation that uses his "life's work" as the technological cornerstone of their global commerce strategy.

3D in Education

Dr. Carroll gave an overview of how ViziTech is using 3D technology to enhance educational techniques in the classroom and other learning venues. His company has been studying the neurological impact of using 3D technology as a tool to engage students on a level virtually unattainable by any other classroom-based methods today. Dr. Carroll described how using 3D models within one's "haptic window" (the area within your immediate reach) stimulates a set of neurons that are not typically engaged using the "tell me-show me" teaching methods. Stimulating these particular neurons apparently improves retention and recall rates that are orders of magnitude better than "tell me-show me." He also provided metrics from cross-industry studies showing quantifiable benefits like mistake-reduction percentage, operational cost savings, and test score improvements.

ViziTech's History

> Founder Stewart Rodeheaver (Brigadier General Retired) has over 40 years experience in the military, business, education, training, and technology fields. He and his team have made it their goal to help raise the bar for educational and training platforms in America.
>
> The company—ViziTech USA—is a breakout learning technology company involved in lead-

ing edge research and application of unique three-dimensional training environments. Our unique new 3D training technologies have been presented to and/or have been purchased by multiple interested government entities, including the Chairman of the Joint Chiefs of Staff, the United States Army Intelligence Center at Ft Huachuca AZ, the DNA and Forensics Training Department at the US Army Criminal Investigation Laboratory, the US State Department, the Georgia Department of Transportation, and others.

We have also installed our unique new 3D training technologies in a growing number of colleges and universities, such as Savannah Technical College, Georgia Regents University, and Embry Riddle University. We have begun actively producing similar training products for fortune 100 corporations such as The Southern Company and Gulfstream Aviation.

Since our beginning, we have received multiple government and civilian awards. Some of these awards include the "Maverick Award" at the 2010 US Army Distance Learning National Conference, and Three prestigious Telly Awards in 2011 for high quality commercial 3D training films. We received from the Technology Association of Georgia / TechAmerica 2011 awards for both "Cool Technology" of the year award (2011) and "Technology Entrepreneur" or the year awards (2011)—beating out IBM, Convergent Technologies, and other super-corporation nominees.[8]

The company is a Georgia based company, but it will have a demonstration site in Alexandria, Virginia.

Meet Some of ViziTech's Clients

Our clients are a reflection of our history. Tasking our super talented crew with meeting the needs of businesses in some of the top technology industries in the world has truly helped push our creativity and technology to the max.

Our Clients Include:

Georgia Institute of Technology

Gulfstream Aviation

Embry Riddle Aeronautical University

Lockheed Aircraft

US Army Crime Lab

New Mexico State Junior College

Georgia Regents University

Georgia Power Company

US Army Biometrics and Forensics Department

Athens Technical College

Westminster Christian Academy, Huntsville, Alabama

University Ga. School for the Deaf

Okefenokee Technical College

Southeastern Technical College

Wiregrass Technical College

Worth county School System

Valdosta City Schools

Colquit County School System

Warner Robins Air Force Base Air Museum

Gwinnett Technical College

Georgia Department of Quick Start Training

Valdosta State University

Atlanta Technical College

Savannah Technical College

Ogeechee Technical College

Albany Technical College

Technical College System of Georgia

Southern Crescent Technical College

SW Ga. Technical College

Moultrie Technical College

Georgia Department of Transportation

Air Safety Corporation

Coffee County School System

Irwin County School System

Pellham City School System

Crisp County School System

Dalton/Whitfield County School System

Griffin/Spalding county School System

Mitchell County School System

Toombs County School System

Charlton County School System

Brantley County School System

Decatur County School System

Murray County School System

US Army, Camp Bullis Texas

We are ready to support your team also with our award winning programs, industry-leading technologies and proven track records of students increased achievements and assessments.

ViziTech's Mission

The Millennium generation we now deal with is considered to be a "Three Screen Generation." By that, we mean that they learn visually, from TV's, Computers, and Cell Phones, which provide multiple visual inputs, often times viewed simultaneously. We have to reach out, capture this method of input that these "digital natives" are accustomed to, and use it to reach the educational or training levels we need them to achieve."[9]

Contact the ViziTech Team and let us go to work for you:

Wellington at HYPERLINK "mailto:wewatts2@gmail.com"wewatts2@gmail.com or HYPERLINK "mailto:wwatts@patriot.net" wwatts@patriot.net

ViziTech's Products: The Hardware

The AV Rover: A Mobile 3D Imaging Projection Center

The AV Rover portable 3D system—The First Fully-Integrated Portable 3D Stereoscopic System—You can now focus on your audience, not the technical complexities behind 3D projection.

The 3D AVRover includes everything you need:

Projector, Computer, Sound System and Software all Configured Properly
Rugged, Portable and Secure Quick Setup and Easy to Move from Room to Room
(Projects on any screen or surface)—Eyewear Storage and Sanitization

Specifications for the AV Rover portable 3D system: CONSOLE—Steel, powder coated, scratch resistant—Dimensions: 36"H x 19"W x 24"D

Tamper-resistant fasteners on all components and panels
Storage rack inside lockable drawer for 3D active glasses
Keyboard and remote storage drawer—Convenient power outlets (2)
25 ft. retractable power cord—Heavy-duty 5" rubber wheels
Removable rear access panel for service

PROJECTOR—Brightness: 2500 ANSI lumens—Native Resolution: XGA

Video Compatibility: NTSC/PAL/SECAM; HDTV (480i, 480p, 720p, 1080i)

Lamp Life: 4000 hours (Economy Mode)—3D Active DLP-Link Glasses
(Quantity to be specified by end user)

3D COMPUTER—2.93 Ghz Intel Core 2 Duo Processor—NVidia 3D Quadro

Graphics Processor—8 GB RAM—500 GB Hard Disk

CD/DVD Player Recorder—WiFi and Ethernet—Windows 7 Pro

MIXER AMPLIFIER—Output: 25 watts RMS—Sources: Computer, laptop, microphone, aux Outputs: 3-Speakers, line-level jacks

WARRANTY: 5 Year AVRover Console and Amplifier—3 Year Projector

1 Year Computer

Vizitech USA is proud to present our first line of classroom tablets. These tablets are fast, reliable, sturdy and offer the student more flexibility than any other tablet available today. Whether you are accessing wireless information, or want to add books, files or data through the USB port or SD card reader, these tablets can offer you the most options.

If you are ready to make the technology jump, the 1G memory with 8G HDD capacity of our

2D Techbook can provide all of the memory and processing strength you expect in an Android version 4.0.4 tablet, and easily handles the amazing graphics.

Each tablet comes with a 1-year warranty. We can provide a second year warranty, and we offer our exclusive "Hot Seat Replacement" program so that broken tablets are replaced within 24 hours. Come join our amazing Tablet Team!!

2D Techbook Droid Tablet

ViziTech Techbook "Droid Tablets–10,1" Android 4.2 ICS Tablets come with 2 USB ports, HDMI connection and an SD card reader. A20 processor, audio port, micro USB port (with adapter for full size), mini HDMI port, Micro SD card, Front and Rear cameras. Comes with a protective case and keyboard all-in-one.

3D Techbook Mini

ViziTech USA "Techbook Mini" 7" Full 3D tablet is smaller than our original Techbook. At 7 inches, this tablet is easier to store into your cases or carry bags. But don't let the small size fool you. The graphics on the Techbook Mini are extremely clean, crisp, and have complete 3D detail. The tablet is pre-programmed to play either 3D pictures or 3D HD Videos, and you can upload as many as you want by using the external Mini SD Card reader built into the case of the tablet. Grab a part of the future and see what 3D with no glasses will do for your visions of Tomorrow—Today![10]

ViziTech Software

National Science Education Standards K–4
3D Video + Interactive Content List

National Science Education Standards K4

3D Video & Interactive Content List

List of 3D Curriculum Videos & Interactives available from ViziTech USA*

*All of the listed videos and interactives come with the AVRover 3D Classroom Package from ViziTech USA

Category	TOTAL TOPIC	TOTAL DURATION
Physical Science	29	01.51.51
Life Science	24	01.01.30
Earth and Space Science	15	01.00.45
Science and Technology	6	00.18.49
Science in Personal and Social Perspectives	3	00.14.39
TOTAL TOPICS	**77**	**04.27.34**

CONTENT STANDARD: K4

	Topic Name	Duration	Interactive
	Physical Science		
1.	Reflection of Light and its Laws	00.04.08	√
2.	Electric Bell	00.03.30	√
3.	Matter and its Properties	00.02.46	
4.	Distinguishing a Physical Change from a Chemical Change	00.03.14	
5.	Floating and Sinking	00.03.42	
6.	Separation of Mixtures using Sublimation	00.01.51	
7.	Separation of Soluble Components from a Mixture	00.02.45	
8.	Molecular Arrangement in Three States of Matter	00.03.16	
9.	The Three Phases of Water	00.02.46	
10.	States of Matter (Part - II)	00.03.16	
11.	Forces and Motion	00.05.16	
12.	Gravity	00.02.33	
13.	Force- A Push or a Pull	00.02.22	
14.	Increasing and Reducing Friction	00.04.54	
15.	Effects of Force	00.03.01	
16.	Sound and Hearing	00.04.48	
17.	Transmission of Sound	00.05.12	
18.	Sound	00.03.52	
19.	Transparent, Translucent, and Opaque Objects	00.02.39	

CONTENT STANDARD: K4

	Topic Name	Duration	Interactive
20.	Colour of Objects	00.05.21	
21.	Thermal Conductors and Insulators	00.02.19	
22.	Heat Energy	00.03.47	
23.	Transformation of Energy	00.08.02	
24.	Transfer of Heat (Conduction)	00.04.12	
25.	Renewable Resources of Energy	00.04.35	
26.	What are Acids and Bases?	00.02.20	
27.	Does Mass Change in a Chemical Reaction?	00.03.34	
28.	Changes accompanying Chemical Reactions	00.03.37	
29.	The common component of all acids	00.08.13	

Life Science

1.	Excretory system	00.03.50	√
2.	Fixed joints and muscles	00.05.17	√
3.	Nervous system	00.03.13	√
4.	Respiratory system	00.03.47	√
5.	Mechanism of breathing in man	00.02.20	√
6.	Respiratory and circulatory systems	00.03.16	√
7.	Digestive and excretory system	00.03.02	√
8.	Parts of a flower	00.01.54	√
9.	Heart- an amazing pump	00.02.43	√
10.	Kidney- The Body Filters	00.02.29	√
11.	Brain- the human computer	00.02.11	√

CONTENT STANDARD: K4

Topic Name		Duration	Interactive
12.	Digestive system	00.02.56	√
13.	Respiration (Breathing)	00.03.37	√
14.	Neurons	00.01.52	√
15.	Plant	00.02.38	√
16.	Muscular system	00.02.24	
17.	Incredible Human Machine (I)	00.02.58	
18.	Food Chain	00.04.00	
19.	Attacking the invaders	00.02.18	
20.	Importance of Forests	00.04.45	
21.	Herbs, shrubs and trees		√
22.	Leaf		√
23.	Axial skeleton (Ribs and sternum)		√
24.	Root		√

Earth and Space Science

		Duration
1.	Rock Cycle	00.03.57
2.	Layers of the Earth's Atmosphere	00.06.00
3.	Day and Night Cycle	00.03.08
4.	Solar Energy	00.05.38
5.	The Sun (Part-1)	00.05.15
6.	The Sun (Part-2)	00.05.25
7.	Chemical Weathering	00.03.14
8.	Earthquake	00.04.13

CONTENT STANDARD: K4

Topic Name	Duration Interactive
9. Erosion	00.03.00
10. Biological Weathering	00.03.11
11. Physical Weathering	00.03.14
12. What must you do during an Earthquake?	00.01.53
13. Plate Tectonics	00.05.46
14. Water Cycle	00.02.36
15. Seasons	00.04.15
Science and Technology	
1. Inclined Plane	00.02.52
2. Lever	00.03.02
3. Pulley	00.03.45
4. Wedges and Screws	00.03.11
5. Oral care	00.01.46
6. Recycling Waste	00.04.13
Science in Personal and Social Perspectives	
1. Rubbish and Litter	00.04.48
2. Saving Energy and the Environment	00.05.57
3. Noise Pollution	00.03.54
TOTAL TOPIC IN CONTENT STANDARD: K4 - 77	**04.27.34**

National Science Education Standards 5–8
3D Video + interactive Content List

National Science Education Standards 5 - 8

3d video + interactive Content List

List Of 3D Curriculum Videos Available From ViziTech USA*

*** All of the listed videos come with the AVRover 3D Classroom Package from ViziTech USA**

Category	TOTAL TOPIC	TOTAL DURATION
Physical Science	121	08.02.26
Life Science	163	08.08.15
Earth and Space Science	7	00.25.26
Science in Personal and Social Perspective	27	01.13.25
Grade 5	7	00.18.23
Grade 6	21	00.47.53
Grade 7	30	01.40.02
Grade 8	14	00.46.02
TOTAL TOPICS	**390**	**21.21.52**

CONTENT STANDARD: 5 - 8

	Topic Name	Duration	Interactive
	Physical Science		
1.	Unit Cells in Crystal Lattices or Space Lattices Part-I	00.08.53	√
2.	Factors Affecting the Solubility of a Solute in a Solvent	00.07.14	√
3.	Osmosis and Osmotic Pressure	00.04.00	√
4.	Laboratory Preparation of Ammonia	00.03.55	√
5.	Preparation of Oxygen from Hydrogen Peroxide	00.01.27	√
6.	Laboratory Preparation of Chlorine	00.02.03	√
7.	Chemical Bonding	00.05.10	√
8.	Flame Test	00.01.47	√
9.	Faraday's First Law of Electrolysis	00.07.17	√
10.	Physical Properties of Alkali Metals	00.03.53	
11.	Physical Properties of Aldehydes and Ketones	00.05.35	
12.	Physical Properties of Amines	00.03.32	
13.	States of Matter	00.03.16	
14.	States of Matter (Part - II)	00.03.16	
15.	Effect of Temperature on the Liquid State of Matter	00.03.36	
16.	Corrosion	00.04.23	
17.	Sublimation	00.01.51	
18.	Crystal Hydrate	00.04.02	

CONTENT STANDARD: 5 - 8

Topic Name		Duration	Interactive
19.	Unit Cells in Crystal Lattices or Space Lattices Part-II	00.05.57	
20.	EDTA Titration Method	00.03.17	
21.	Chemical Reactions and their Characteristics	00.02.50	
22.	Chemical Equations	00.04.00	
23.	The Law of Conservation of Mass	00.01.26	
24.	Chemical Properties of Carbon	00.02.19	
25.	Chemical Properties of Sulphur Dioxide - I	00.06.10	
26.	Chemical Properties of Sulphur Dioxide-II	00.06.31	
27.	Chemical Properties of Dilute Sulphuric Acid	00.02.20	
28.	Chemical Properties of Phosphorus	00.03.34	
29.	Chemical Properties of Metals and Non metals (Reaction with oxygen)	00.03.37	
30.	Chemical Properties of Phenol	00.08.13	
31.	Molecular Size and Solubility	00.02.02	
32.	Factors Affecting Vapour Pressure of a Solvent	00.02.57	
33.	Vapour Pressure of a Solvent	00.01.57	
34.	Effect of Temperature On the Solubility of Gases and Thermal Pollution	00.04.49	
35.	Physical Properties of Non-metals	00.04.21	
36.	Laboratory Preparation and Physical Properties of Phenol	00.03.25	
37.	Sulphur	00.02.22	
38.	Laboratory Preparation of Sulphur Dioxide	00.03.04	

CONTENT STANDARD: 5 - 8

	Topic Name	Duration	Interactive
39.	Preparation of Carbon Dioxide	00.01.51	
40.	Laboratory preparation and detection of nitric acid	00.05.11	
41.	Allotropic forms of Phosphorus	00.03.59	
42.	Reaction of Metals with Oxygen	00.04.46	
43.	Galvanization	00.07.20	
44.	Applications of Electrolysis (Part-I)	00.04.06	
45.	Standard Hydrogen Electrode	00.01.37	
46.	Use of Standard Hydrogen Electrode as an Anode	00.05.59	
47.	Use of Standard Hydrogen Electrode as a Cathode	00.05.00	
48.	Electrode Potential and Standard Electrode Potential	00.04.41	
49.	Faraday's Second Law of Electrolysis	00.02.41	
50.	Heat of Solution	00.03.57	
51.	Concentration(Molarity) of a Solution		√
52.	Molecular Library		√
53.	States of Matter		√
54.	Density of a Cubic Crystalline Solid		√
55.	Introduction to Modern Periodic Table		√
56.	Geometry of $[Fe(H_2O)_6]^{3+}$		√
57.	Geometry of $[Co(NH_3)_6]^{2+}$		√
58.	Geometry of $[K_3Fe(CN)_6]$		√

CONTENT STANDARD: 5 - 8

Topic Name		Duration	Interactive
59.	Law of Mass Action		√
60.	The Atomic Spectra		√
61.	Maxwell Speed Distribution		√
62.	Electronegativity and bond character-Part-2		√
63.	Dynamic nature of chemical equlilibrium		√
64.	Density (Part-1)	00.05.09	√
65.	Acceleration	00.08.49	√
66.	Thermal Conductivity	00.06.31	√
67.	Internal Combustion Engine	00.06.07	√
68.	Reflection of Light and its Laws	00.04.08	√
69.	Curved Mirrors	00.03.24	√
70.	Images Formed by a Convex Mirror	00.04.42	√
71.	Images Formed by a Concave Mirror	00.05.38	√
72.	Dispersion of White Light	00.02.47	√
73.	Total Internal Reflection (Part-2)	00.04.01	√
74.	Lenses	00.05.20	√
75.	Converging Lenses	00.07.36	√
76.	Diverging Lenses	00.06.27	√
77.	Compound Microscope	00.04.40	√
78.	Galvanometer	00.09.05	√
79.	Fundamentals of a DC Motor	00.07.31	√
80.	Solar Energy (Active Solar Systems)	00.04.48	√
81.	Latitude and Longitude	00.07.17	√

CONTENT STANDARD: 5 - 8

Topic Name		Duration	Interactive
82.	Volume	00.02.23	
83.	Volume of an Irregular Object (Part-1)	00.01.30	
84.	Archimedes' Principle	00.06.38	
85.	Inertia	00.03.48	
86.	Newton's First Law of Motion	00.03.11	
87.	Effects of Force	00.03.01	
88.	Rolling Friction and Sliding Friction	00.04.10	
89.	Increasing and Reducing Friction	00.04.54	
90.	Balanced Forces	00.04.24	
91.	Kinetic Energy	00.05.16	
92.	Transformation of Energy	00.08.02	
93.	Transfer of Heat (Conduction)	00.04.12	
94.	Kinetic Theory of Matter-1 (Solid)	00.05.27	
95.	Sound	00.03.52	
96.	Infrasonics and Ultrasonics	00.05.13	
97.	Transparent, Translucent, and Opaque Objects	00.02.39	
98.	Colour of Objects	00.05.21	
99.	Atmospheric Refraction	00.07.03	
100.	Periscope	00.05.43	
101.	Real and Virtual Images	00.07.20	

CONTENT STANDARD: 5 - 8

Topic Name	Duration	Interactive
102. Sign Convention for Spherical Mirrors and Lenses	00.06.10	
103. Lasers and their Uses	00.06.35	
104. Applications of Laser (Part-2) (Barcode Reader)	00.02.19	
105. Projector	00.06.37	
106. Fresnel lens and Overhead Projector	00.05.51	
107. Prism Spectrometer	00.03.55	
108. Working of a Prism Spectrometer	00.07.06	
109. Diffraction of Light	00.06.51	
110. Capacitor	00.0[illegible]	
111. Electroplating	00.02.56	
112. Electric Fuse	00.03.19	
113. Solar Energy (Passive Solar Heating and Photovoltaic Devices)	00.05.55	
114. The Sun (Part-1)	00.05.15	
115. The Sun (Part-2)	00.05.25	
116. Thermal Radiation	00.03.15	
117. Motion of a Block on an Inclined Plane		√
118. Density of Fluids		√
119. Rotation of a Plane Mirror		√
120. Volume Expansion of Solids		√

CONTENT STANDARD: 5 - 8

	Topic Name	Duration	Interactive
	Life Science		
1.	Water potential	00.02.47	√
2.	Typical flower	00.04.01	√
3.	Support in aquatic plants	00.01.11	√
4.	Brassicaceae (Mustard plant)	00.03.34	√
5.	Structure of mitochondria	00.02.38	√
6.	Mitosis	00.03.10	√
7.	Ponds & lakes	00.02.07	√
8.	Chromosomes, Genes and DNA	00.03.50	√
9.	Types of muscle fibers	00.04.24	√
10.	The skull	00.01.20	√
11.	Vertebral column	00.01.36	√
12.	Sternum and ribs	00.01.51	√
13.	Bone disorder (Osteoporosis)	00.03.06	√
14.	Fractures	00.02.46	√
15.	Atherosclerosis (Heart disease)	00.02.04	√
16.	Disorders of the heart	00.02.13	√
17.	Structure of neuron	00.03.45	√
18.	Structure of antibody	00.03.02	√
19.	Chicken pox	00.03.36	√
20.	Influenza and bird flu	00.03.17	√
21.	Polio	00.01.52	√
22.	Phylum Cnidaria	00.02.40	√

CONTENT STANDARD: 5 - 8

Topic Name		Duration	Interactive
23.	Viruses	00.05.08	√
24.	Virus (Bacteriophage)	00.01.09	√
25.	Amoeba	00.01.16	√
26.	Algae	00.04.33	√
27.	Cockroach	00.03.01	√
28.	Excretory system (Cockroach)	00.02.38	√
29.	Aphids	00.01.27	√
30.	Algae and diatoms	00.02.59	√
31.	Yeast and fungi	00.03.29	√
32.	AIDS	00.04.06	√
33.	Shape of cells	00.04.37	√
34.	Monocot root anatomy	00.03.06	√
35.	Female reproductive anatomy	00.03.29	√
36.	Ultrastructure of skeletal muscles	00.03.18	√
37.	Diarrhoea	00.03.12	√
38.	Floral arrangement	00.05.17	√
39.	Fish	00.04.10	√
40.	Mechanism of muscle fibre contraction	00.03.59	√
41.	Joints	00.03.14	√
42.	The plant	00.03.05	√
43.	Solanaceae	00.02.52	√
44.	Ecological pyramids	00.05.07	√
45.	Transpiration	00.03.10	√

CONTENT STANDARD: 5 - 8

	Topic Name	Duration	Interactive
46.	Immunity	00.04.21	√
47.	Glucose metabolism disorder (Diabetes)	00.10.41	√
48.	Complications of Diabetes	00.03.46	√
49.	Diabetes- It's kinds	00.03.55	√
50.	Physiology of digestion	00.04.31	√
51.	Mobility of human gut	00.04.00	√
52.	Bones	00.03.45	√
53.	Anatomy of the bone	00.03.00	√
54.	Fixed joints and muscles	00.05.17	√
55.	Respiratory system (Larynx)	00.06.09	√
56.	Heart and Blood circulation	00.03.35	√
57.	Exocrine glands and Endocrine glands	00.04.49	√
58.	Thyroid gland	00.03.56	√
59.	Nephron (Structure & functions)	00.02.55	√
60.	Ultrastructure of kidney	00.02.37	√
61.	Skin	00.03.08	√
62.	Appendicular skeleton	00.04.35	√
63.	Girdle bones	00.02.46	√
64.	Mechanism of hearing	00.05.13	√
65.	Eye (Anatomy and physiology)	00.05.47	√
66.	Tongue - organ of taste	00.04.17	√
67.	Mechanism of breathing in man	00.02.20	√
68.	Nervous system	00.03.13	√

CONTENT STANDARD: 5 - 8

	Topic Name	Duration	Interactive
69.	Respiratory and circulatory systems	00.03.16	√
70.	Respiratory system	00.03.47	√
71.	Gaseous transport	00.06.33	√
72.	Human heart (I)	00.04.32	√
73.	Heart	00.02.07	√
74.	Male reproductive system (Anatomy & physiology)	00.03.56	√
75.	Arteries and veins	00.05.19	√
76.	Leucocytes (Structure and function)	00.02.34	√
77.	Blood corpuscles (Human blood)	00.03.24	√
78.	Thermoregulation by the skin	00.03.51	√
79.	Accessory digestive organs	00.03.16	√
80.	Digestive and excretory system	00.03.02	√
81.	Detailed digestive system II	00.01.52	√
82.	Eukaryotic cell	00.03.03	√
83.	Parts of a flower	00.01.54	√
84.	Morphology of flower	00.04.00	√
85.	Malvaceae	00.02.57	√
86.	Cell wall and cell structure (Plant cell)	00.02.32	√
87.	Brain (Anatomy & function)	00.04.37	√
88.	Excretory system	00.03.50	√
89.	Connective tissues	00.02.07	
90.	Epithelial tissue	00.04.31	

CONTENT STANDARD: 5 - 8

Topic Name	Duration	Interactive
91. Areolar connective tissues	00.03.27	
92. Plant tissues (Meristematic tissues)	00.01.58	
93. Plant tissue system	00.04.11	
94. Simple tissues	00.02.56	
95. Muscular system	00.02.41	
96. Food chain	00.02.52	
97. Forest ecosystem	00.02.12	
98. Respiration in lower animals (Hydra and cockroach)	00.05.39	
99. Digestive system of cockroach	00.02.39	
100. Respiration in plants(During the night)	00.01.42	
101. Homeostatic Regulation (Kidney)	00.04.16	
102. Gall bladder	00.03.03	
103. Cell structure III	00.01.15	
104. Structure of onion peel and cork cell	00.01.34	
105. Number & size of cells	00.03.38	
106. Meiosis	00.06.12	
107. Bulk transport	00.03.12	
108. Leaf of maize	00.03.54	
109. Fruits	00.04.45	
110. Inflorescence I	00.07.26	
111. Inflorescence II	00.03.18	
112. Liver	00.05.02	

CONTENT STANDARD: 5 - 8

Topic Name	Duration	Interactive
113. Role of Muscles In Bone Movement	00.04.09	
114. Types of pathogenic bacteria	00.03.14	
115. Deficiency diseases	00.01.24	
116. Caffeine addiction and effects	00.03.49	
117. Disorders of the brain (Multiple sclerosis)	00.03.41	
118. Mycoplasma	00.02.10	
119. Disorders of the brain (Alzheimer's disease)	00.02.20	
120. Passive smoking and bronchitis	00.03.20	
121. Asthma	00.03.26	
122. Haemoglobin and sickle cell anaemia	00.04.19	
123. Oogenesis	00.05.17	
124. Spermatogenesis in human beings	00.04.02	
125. Fertilization and implantation in humans	00.03.44	
126. Placenta & foetal haemoglobin	00.04.09	
127. Bacterial transformation and conjugation	00.02.47	
128. Reflex arc	00.03.03	
129. Regeneration among animals	00.01.46	
130. Support system in aquatic & terrestrial plants	00.02.30	
131. Plasmolysis	00.03.00	
132. Hydroponics	00.03.33	
133. Energy flow in an ecosystem	00.03.48	
134. General characteristics of insects	00.05.05	
135. Worms	00.02.49	

CONTENT STANDARD: 5 - 8

Topic Name	Duration	Interactive
136. Osteichthyes and chondrichthyes	00.05.23	
137. Starfish and snail	00.03.17	
138. Liliaceae	00.02.22	
139. Locomotion in animals	00.02.48	
140 Filter feeding	00.02.26	
141 Conditions required for seed germination	00.01.54	
142. Axial skeleton (Ribs and sternum)		√
143. Difference between vertebrates and invertebrates		√
144. Functions of Golgi and ER		√
145. Gaseous exchange in animals		√
146. Ligaments		√
147. Prokaryota and eukaryota		√
148. Prokaryotic Cell		√
149. Respiratory System		√
150. Types of flowers		√
151. Blood Clotting		√
152. Thermoregulation by Skin		√
153. Types of fractures		√
154. Viral Replication		√
155. DNA Replication		√
156. Food web		√
157. Insectivory		√

CONTENT STANDARD: 5 - 8

Topic Name		Duration	Interactive
158.	Adaptation of leaves for photosynthesis		√
159.	Rhodophyceae		√
160.	Alternation of generation in Spirogyra		√
161.	Fabaceae (Pisum sativum)		√
162	Root		√
163	Anatomy of root		√
	Earth and Space Science		
1.	Kepler's Second Law of Planetary Motion	00.06.00	√
2.	Solar Energy (Active Solar Systems)	00.04.48	√
3.	Solar System	00.01.57	
4.	Solar Energy (Passive Solar Heating and Photovoltaic Devices)	00.05.55	
5.	Lunar Eclipse	00.05.26	
6.	Magnetic Declination	00.01.20	
7.	Reflecting Telescope		√
	Science in Personal and Social Perspective		
1.	Darkfield Microscopy	00.02.00	√
2.	Gene amplification using polymerase chain reaction (PCR)	00.02.47	√
3.	Industrial production of enzymes	00.04.09	√
4.	Monoclonal antibodies	00.03.33	√

CONTENT STANDARD: 5 - 8

	Topic Name	Duration	Interactive
5.	Carbohydrates	00.01.49	√
6.	Fats	00.01.05	√
7.	Simple test for carbohydrate, fats and protein	00.02.18	√
8.	Chicken pox	00.03.36	√
9.	Influenza and bird flu	00.03.17	√
10.	AIDS	00.04.06	√
11.	Tissue culture	00.04.19	√
12.	Cancer therapy (Nanotechnology)	00.04.54	√
13.	Oral care	00.01.46	√
14.	Application of genetic engineering	00.03.15	√
15.	ECG and EEG	00.04.41	
16.	Types of vaccines	00.02.11	
17.	Deficiency diseases	00.01.24	
18.	Caffeine addiction and effects	00.03.49	
19.	Somatic hybridization	00.02.47	
20.	Air pollution	00.02.57	
21.	High calorie diet and obesity	00.03.09	
22.	Effect of antibiotics	00.02.44	
23.	Smoking and Emphysema	00.04.21	
24.	Introductory Microscope Experiments	00.02.28	
25.	Sphygmomanometer		√
26.	Malaria		√
27.	AIDS		√

CONTENT STANDARD: 5 - 8

	Topic Name	Duration	Interactive
	Grade 5		
1.	Basic concepts of Decimal	00.03.33	√
2.	Concept of Fraction	00.05.09	√
3.	Representation of a Decimal Number on the Number Line	00.03.30	
4.	Fractions to Decimals and Decimals to Fractions	00.06.11	
5.	Equivalent Fraction		√
6.	Properties of Whole Numbers (Multiplication)		√
7.	Volume		√
	Grade 6		
1.	Ratio	00.03.34	√
2.	Prime Numbers	00.03.29	√
3.	Factors of a Number	00.03.22	√
4.	Areas of Plane Figures	00.05.53	√
5.	Volume of a Prism	00.04.06	√
6.	Regular Octahedron	00.02.30	√
7.	Tally Marks	00.04.04	√
8.	Introduction to Bar graph	00.04.21	√
9.	Pie Chart	00.03.35	√
10.	HCF / GCF	00.03.31	

CONTENT STANDARD: 5 - 8

	Topic Name	Duration	Interactive
11.	Introduction to Integers	00.03.28	
12.	Introduction to Graphs	00.06.00	
13.	Walk on Multiples		√
14.	Least Common Multiple		√
15.	Walk on Composites		√
16.	Multiplication of Integers		√
17.	Addition of Integers		√
18.	Classification of Quadrilaterals		√
19.	Concept of Area		√
20.	Concept of Perimeter		√
21.	Ascending and Descending Order		√

Grade 7

	Topic Name	Duration	Interactive
1.	Ratio	00.03.34	√
2.	Proportion	00.04.04	√
3.	Direct & Inverse Proportion	00.03.51	√
4.	Concept of Fraction	00.05.09	√
5.	Introduction to Algebra	00.03.41	√
6.	Construction: Angle Bisector and an angle of 30°	00.02.44	√
7.	Construction: An Angle of 60° and 120°	00.03.22	√
8.	Construction: Right Angled Triangle	00.03.17	√
9.	Types of a Triangle	00.04.01	√

CONTENT STANDARD: 5 - 8

Topic Name		Duration	Interactive
10.	Introduction to Prism	00.05.31	√
11.	Introduction to Pyramid	00.04.31	√
12.	Area of a Circle	00.02.39	√
13.	Definition of pi (π)	00.03.14	√
14.	Angle	00.03.47	√
15.	Types of Angles formed by a Transversal	00.04.15	√
16.	Surface Area of Rectangular Solids	00.04.35	√
17.	Surface Area of a Sphere	00.04.19	√
18.	Surface Area of a Right Prism	00.02.54	√
19.	Surface Area of a Pyramid	00.02.12	√
20.	Surface Area of a Cone	00.04.18	√
21.	Volume of a Prism	00.04.06	√
22.	Areas of Plane Figures	00.05.53	√
23.	Probability	00.02.57	√
24.	Important Points associated with a Triangle	00.05.08	
25.	Lines and Angles	00.03.46	
26.	Area of the Frame	00.02.14	
27.	Properties of Whole Numbers (Multiplication)		√
28.	Volume		√
29.	Concept of Area		√
30.	Surface Area of a Cylinder		√

CONTENT STANDARD: 5 - 8

	Topic Name	Duration	Interactive
	Grade 8		
1.	Basic concepts of Decimal	00.03.33	√
2.	Solving an equation in one variable (Balancing method)	00.04.08	√
3.	Constructions: Line segment, Circle & Perpendicular Bisector	00.03.09	√
4.	Angle	00.03.47	√
5.	Types of Angles formed by a Transversal	00.04.15	√
6.	Pythagoras' theorem	00.02.39	√
7.	Volume of a Cone	00.02.56	√
8.	Volume of a Cylinder	00.04.25	√
9.	Finding Cube Roots	00.04.21	
10.	Function	00.05.32	
11.	Measurement of Line segment and Angle	00.03.31	
12.	Lines and Angles	00.03.46	
13.	Angle Sum Property of Triangle		√
14.	Exterior angle of a triangle		√

TOTAL TOPIC IN CONTENT STANDARD: 5 - 8 - 390 21.21.52

National Science Education Standards 9–12
3D video + Interactive Content List

Legendary Products, Tomorrow's Solutions

National Science Education Standards 9 - 12

3D Video & Interactive Content List

List of 3D Curriculum Video & Interactive available from ViziTech USA*

*All of the listed videos and interactives come with the AVRover 3D Classroom Package from ViziTech USA

Category	TOTAL TOPIC	TOTAL DURATION
Physical Science	200	14.09.48
Life Science	115	05.14.06
Earth And Space Science	8	00.20.32
Science in Personal and Social Perspective	18	00.44.48
High School-Algebra	10	00.25.59
High School-Functions	11	00.40.03
High School-Geometry	46	02.23.47
High School-Number and Quantity	6	00.19.59
High School-Statistics and Probability	2	00.08.03
TOTAL TOPICS	**416**	**24.27.05**

CONTENT STANDARD: 9 - 12

	Topic Name	Duration	Interactive
	Physical Science		
1.	Molecular Formula	00.02.22	√
2.	Chemical Bonding	00.05.10	√
3.	The Sigma (σ) and the Pi (π) Bonds	00.07.25	√
4.	Bond Energy	00.07.29	√
5.	Electronegativity	00.04.39	√
6.	Atomic Radius and its Types	00.02.52	√
7.	Periodic Trends in Atomic Radii	00.02.39	√
8.	Ionization Energy	00.02.54	√
9.	Hybridization of Atomic Orbitals	00.03.11	√
10.	Electronic Configuration and the Position of an Element in the Periodic Table	00.05.34	√
11.	The Mole Concept	00.04.14	√
12.	Rules for Filling of Orbitals	00.07.09	√
13.	Quantum Numbers (Part -I)	00.06.11	√
14.	Sulphonation of Benzene	00.02.15	√
15.	Dipole Moment	00.05.31	√
16.	Number of Particles in a Cubic Unit Cell	00.06.32	√
17.	Difference Between Diamond and Graphite	00.04.46	√
18.	Nomenclature of Aliphatic Amines	00.06.32	√
19.	IUPAC Nomenclature of Acyclic Organic compounds	00.06.02	√
20.	Alkenes	00.06.04	√
21.	Application and Explanation of Henry's Law	00.06.08	√

CONTENT STANDARD: 9 - 12

Topic Name		Duration	Interactive
22.	Temperature Units and their Inter-conversion	00.04.13	√
23.	Effect of Temperature on the Liquid State of Matter	00.03.36	
24.	Emission Spectra and the Flame Test	00.06.02	
25.	Inductive Effect	00.08.11	
26.	Electromeric Effect	00.06.31	
27.	Introduction to Valence Bond Theory	00.02.29	
28.	VSEPR Theory	00.04.02	
29.	Law of Definite Proportions or Constant Composition	00.04.50	
30.	Law of Multiple Proportions	00.05.00	
31.	Balancing of Equations	00.02.40	
32.	Valency	00.06.58	
33.	Covalent Bonds	00.05.16	
34.	Co-ordinate Bond	00.03.19	
35.	Bonding in Carbon	00.05.05	
36.	Properties of Non-Metals Favouring the Formation of Covalent Bonds	00.06.42	
37.	Orbits and Orbitals	00.06.18	
38.	Factors Affecting Atomic Radii	00.05.19	
39.	Factors Favouring the Formation of Ionic Compounds-I	00.05.37	
40.	Factors Favouring the Formation of Ionic Compounds-II	00.05.41	
41.	Coordination Number and Geometry of Complexes	00.02.41	

CONTENT STANDARD: 9 - 12

	Topic Name	Duration	Interactive
42.	Classification of Ligands	00.06.05	
43.	Introduction to Coordination Compounds	00.04.32	
44.	$sp^3 d^2$ Hybridization in SF_6	00.01.27	
45.	Types of Hybridization in Organic Compounds (sp3 Hybridization in Ethane)	00.04.37	
46.	Geometry of K_4 $[Fe(CN)_6]$ based on its Magnetic Property	00.03.02	
47.	Relation between Geometry and Magnetic Properties of Transition metal Complexes	00.03.30	
48.	The Stability of an Expanded Octet of PCl_5	00.04.35	
49.	Atomic Mass, Molecular Mass and Formula Unit Mass	00.06.42	
50.	Mole Fraction	00.03.04	
51.	Formulae of Compounds (Using Valencies)	00.02.48	
52.	Quantum Numbers (Part -II)	00.09.29	
53.	The Empirical Formula and Molecular Formula of a Compound Part-1	00.04.39	
54.	The Empirical Formula and Molecular Formula of a Compound Part-2	00.05.40	
55.	Simple Cubic Crystal Lattices (AAA Arrangement)	00.03.15	
56.	Heat of Hydrogenation and Resonance in Benzene	00.06.12	
57.	Chlorination of Benzene	00.01.56	
58.	The Boiling Points of Alcohols and Hydrogen Bonding	00.05.12	
59.	Dipole Moment and Bond Character	00.03.02	

CONTENT STANDARD: 9 - 12

	Topic Name	Duration	Interactive
60.	Unit Cells in Crystal Lattices or Space Lattices Part-II	00.05.57	
61.	Carbon	00.01.54	
62.	Fullerenes	00.01.25	
63.	Ionic Addition Polymerization	00.03.56	
64.	Tacticity of Polymers	00.03.16	
65.	Addition Reactions of Alkenes	00.03.20	
66.	Diene Compounds	00.01.32	
67.	IUPAC Nomenclature-I(Alkanes)	00.02.09	
68.	Condensation Reactions of Aldehydes and Ketones	00.07.04	
69.	Preparation of Amines	00.04.46	
70.	Reactions of Amines with Nitrous Acid	00.03.04	
71.	Tests to Detect Amines	00.06.31	
72.	Azo Coupling Reaction	00.04.41	
73.	Structural Isomerism in Alkanes	00.03.37	
74.	Nomenclature of Geometric Isomers	00.02.26	
75.	Geometrical Isomerism in Hydrocarbons	00.05.39	
76.	Classification of Functional Groups (Part - II)	00.06.27	
77.	pH Scale and its Limitations	00.08.23	
78.	Destructive Distillation of Coal	00.02.49	
79.	Physical and Chemical Changes	00.03.14	
80.	Types of Organic Reactions	00.05.40	
81.	Mechanism of Nucleophilic Substitution Reaction	00.07.37	

CONTENT STANDARD: 9 - 12

Topic Name	Duration	Interactive
82. Redox Reaction	00.05.34	
83. Polyatomic Ions	00.02.37	
84. Acid Base Titration	00.06.10	
85. Neutralization	00.03.14	
86. Oxidation of Aldehydes	00.07.24	
87. Oxidation of Ketones	00.03.32	
88. Electrophilic and Nucleophilic Reagents	00.05.18	
89. Average and Instantaneous Rate of Reaction	00.07.00	
90. Partial Pressure	00.04.47	
91. Heterogeneous Catalysts	00.03.19	
92. Adsorption	00.07.26	
93. Types of Adsorption	00.02.21	
94. Collision Theory	00.07.19	
95. Diffusion	00.02.50	
96. Calorific Value and Fuel Efficiency	00.04.25	
97. Nickel-Cadmium Battery	00.08.32	
98. The Common Component of all Acids		√
99. Carbonyl Compounds		√
100. Classification of Carbon and Hydrogen Atoms in Alkanes		√
101. Laboratory Preparation of Hydrogen		√
102. Laboratory Preparation of Hydrogen(Quantitative Aspect)		√
103. Limiting Reactant		√
104. Nomenclature of Alcohols		√

CONTENT STANDARD: 9 - 12

Topic Name	Duration	Interactive
105. Structure of Benzene		√
106. Friedel-Crafts Acylation of Benzene		√
107. Friedel-Crafts Alkylation of Benzene		√
108. Methane		√
109. IUPAC Nomenclature of Coordination Compounds		√
110. Nitration of Benzene		√
111. Nomenclature of Aldehydes and Ketones		√
112. Nomenclature of Carboxylic Acid		√
113. Combustion Analysis (Liebig's method)		√
114. The Dry cell		√
115. Heat Conduction and Steady State	00.04.38	√
116. Ripple Tank Experiment to Illustrate Interference	00.06.59	√
117. Direct and Alternating Current	00.07.14	√
118. Ohm's Law	00.07.26	√
119. Magnetic Field due to a Straight Wire Carrying Current (Part-1)	00.06.02	√
120. Magnetic Field due to a Straight Wire Carrying Current (Part-2)	00.03.09	√
121. Electric Bell	00.03.30	√
122. Reed Switch	00.03.29	√
123. Motion of Charge in a Magnetic Field (Part-1)	00.04.17	√
124. Motion of Charge in a Magnetic Field (Part-2)	00.05.23	√
125. Fleming's Left Hand Rule	00.02.01	√

CONTENT STANDARD: 9 - 12

Topic Name	Duration	Interactive
126. Magnetic Dipole Moment	00.07.05	√
127. Torque on a Current Carrying Loop in a Uniform Magnetic Field	00.06.41	√
128. Fundamentals of a DC Motor	00.07.31	√
129. LC Oscillations	00.06.50	√
130. RL Circuit (Growth Phase)	00.08.06	√
131. RL Circuit (Decay Phase)	00.06.01	√
132. Rutherford's Atomic Model	00.04.02	√
133. Size of Nucleus and Nuclear Density	00.05.02	√
134. Discharge of Electricity through Gases	00.02.34	
135. Thermionic Emission	00.08.19	
136. Properties of Nucleons	00.06.26	
137. Radioactivity	00.07.44	
138. Alpha Emission (Part-1)	00.05.03	
139. Beta Emission (Part-1)	00.06.55	
140. Kinetic Theory of Matter-1 (Solid)	00.05.27	
141. Kinetic Theory of Matter-2 (Liquid)	00.04.13	
142. Fluids in Motion- Equation of Continuity	00.09.34	
143. Equation of Continuity (Applications)	00.03.16	
144. Bernoulli's Principle	00.06.24	
145. Applications of Bernoulli's Principle	00.04.38	
146. Capillarity	00.08.49	
147. Force- A Push or a Pull	00.02.22	
148. Electrostatic Force (A Non-contact Force)	00.04.49	

CONTENT STANDARD: 9 - 12

Topic Name	Duration	Interactive
149. Effects of Force	00.03.01	
150. Distance and Displacement	00.05.10	
151. Newton's First Law of Motion	00.03.11	
152. Newton's Second Law of Motion	00.07.07	
153. Newton's Third Law of Motion	00.04.09	
154. Mass and Weight	00.05.05	
155. Moment of a Force and the Law of Moments	00.07.39	
156. Couple	00.07.21	
157. Properties of Electric Charge	00.05.29	
158. Electric Field	00.04.02	
159. Electric Field Lines	00.02.56	
160. Gauss's Theorem (Part-1)	00.04.10	
161. Gauss's Theorem (Part-2)	00.03.59	
162. Application of Gauss's Theorem (Part-1)	00.03.53	
163. Van de Graaff Generator	00.04.26	
164. Electric Current	00.05.40	
165. RC Circuit	00.04.54	
166. Types of Magnetism	00.02.53	
167. Lenz's Law	00.04.13	
168. Transformers	00.09.18	
169. Energy Losses in a Transformer	00.06.20	
170. Transformation of Energy	00.08.02	
171. Transfer of Heat (Conduction)	00.04.12	
172. Kinetic Energy	00.05.16	

CONTENT STANDARD: 9 - 12

Topic Name	Duration	Interactive
173. Renewable Resources of Energy	00.04.35	
174. Sound	00.03.52	
175. Propagation of Sound Waves through Different Media	00.02.47	
176. Doppler Effect in Sound	00.06.11	
177. Different Cases of Doppler Effect	00.05.32	
178. The Sun (Part-1)	00.05.15	
179. The Sun (Part-2)	00.05.25	
180. Electromagnetic Spectrum (Part - 1)	00.05.19	
181. Electromagnetic Spectrum (Part - 2)	00.03.51	
182. X - ray Spectrum	00.06.43	
183. Thermal Radiation	00.03.15	
184. Crystalline and Amorphous Solids	00.05.49	
185. Semiconductors	00.07.05	
186. Intrinsic Semiconductors	00.03.42	
187. Extrinsic Semiconductor (Part-2)	00.02.51	
188. Area Expansion in Solids		√
189. Introduction to Vectors		√
190. Geometrical Representation of Simple Harmonic Motion		√
191. Centripetal Force		√
192. Magnetic Field around a Current Carrying Circular Coil		√
193. Force between Two Parallel Wires Carrying Current		√

CONTENT STANDARD: 9 - 12

Topic Name		Duration	Interactive
194.	Domain Theory of Magnetism		√
195.	Faraday's Law of Electromagnetic Induction		√
196.	AC Generator		√
197.	DC Generator		√
198.	Linear Polarization of Light by Selective Absorption		√
199.	Cathode Ray Tube		√
200.	Frequency Modulation		√

Life Science

1.	Support in aquatic plants	00.01.11	√
2.	Brassicaceae (Mustard plant)	00.03.34	√
3.	Structure of the leaf	00.06.21	√
4.	Structure of mitochondria	00.02.38	√
5.	Synthesis of m-RNA	00.03.29	√
6.	Protein synthesis	00.05.53	√
7.	Mitosis	00.03.10	√
8.	Chromosomes, Genes and DNA	00.03.50	√
9.	Structure of DNA	00.02.55	√
10.	Chromatin structure	00.01.40	√
11.	Mutation	00.02.56	√
12.	Structure of neuron	00.03.45	√
13.	Types of neurons	00.05.28	√
14.	Neuroglial cells	00.02.41	√

CONTENT STANDARD: 9 - 12

Topic Name		Duration	Interactive
15.	Neuromuscular junction	00.02.52	√
16.	Phylum Cnidaria	00.02.40	√
17.	Virus (Bacteriophage)	00.01.09	√
18.	Amoeba	00.01.16	√
19.	Algae	00.04.33	√
20.	Cockroach	00.03.01	√
21.	Algae and diatoms	00.02.59	√
22.	Yeast and fungi	00.03.29	√
23.	Impulse transmission	00.02.57	√
24.	Internal structure of dicot root	00.03.11	√
25.	Fish	00.04.10	√
26.	The plant	00.03.05	√
27.	Solanaceae	00.02.52	√
28.	Ecological pyramids	00.05.07	√
29.	Immunity	00.04.21	√
30.	Physiology of digestion	00.04.31	√
31.	Mobility of human gut	00.04.00	√
32.	Physiology of cell membrane	00.05.06	√
33.	ATP synthesis in mitochondrion	00.04.06	√
34.	DNA structure	00.02.54	√
35.	Karyotype	00.03.32	√
36.	Ribonucleic acid	00.05.37	√
37.	Impulse transmission (Action potential)	00.03.54	√
38.	Exocrine glands and Endocrine glands	00.04.49	√

CONTENT STANDARD: 9 - 12

	Topic Name	Duration	Interactive
39.	Urine formation	00.06.13	√
40.	Mechanism of hearing	00.05.13	√
41.	Eye (Anatomy and physiology)	00.05.47	√
42.	Tongue - organ of taste	00.04.17	√
43.	Spinal cord (Myelon)	00.03.31	√
44.	Nervous system	00.03.13	√
45.	Respiratory system	00.03.47	√
46.	Respiratory and circulatory systems	00.03.16	√
47.	Heart	00.02.07	√
48.	Arteries and veins	00.05.19	√
49.	Leucocytes (Structure and function)	00.02.34	√
50.	Blood corpuscles (Human blood)	00.03.24	√
51.	Oral care	00.01.46	√
52.	Photosynthesis in plants	00.03.49	√
53.	Carbohydrates	00.01.49	√
54.	Biological molecules (Carbohydrates)	00.05.50	√
55.	Thermoregulation by the skin	00.03.51	√
56.	Detailed digestive system II	00.01.52	√
57.	Accessory digestive organs	00.03.16	√
58.	Digestive and excretory system	00.03.02	√
59.	Generalised eukaryotic cell	00.03.43	√
60.	Eukaryotic cell	00.03.03	√
61.	Malvaceae	00.02.57	√
62.	Enzymes as biocatalysts	00.03.22	√

CONTENT STANDARD: 9 - 12

Topic Name	Duration	Interactive
63. Brain (Anatomy & function)	00.04.37	√
64. Excretory system	00.03.50	√
65. The role of ATP in active transport	00.02.18	
66. Extracellular and intracellular enzymes	00.02.56	
67. Bulk transport	00.03.12	
68. Meiosis	00.06.12	
69. Role of insulin in cell metabolism	00.02.13	
70. Food chain	00.02.52	
71. Connective tissues	00.02.07	
72. Epithelial tissue	00.04.31	
73. Areolar connective tissues	00.03.27	
74. Bacterial transformation and conjugation	00.02.47	
75. Euglena	00.04.33	
76. General characteristics of insects	00.05.05	
77. Worms	00.02.49	
78. Osteichthyes and chondrichthyes	00.05.23	
79. Starfish and snail	00.03.17	
80. Liliaceae	00.02.22	
81. Hydroponics	00.03.33	
82. Energy flow in an ecosystem	00.03.48	
83. Air pollution	00.02.57	
84. Physiology of the photoperiodism	00.02.56	
85. Liver	00.05.02	
86. Muscular system	00.02.41	

CONTENT STANDARD: 9 - 12

	Topic Name	Duration	Interactive
87.	Reflex arc	00.03.03	
88.	The Olfactory system of human beings	00.02.52	
89.	Herbs, shrubs and trees		√
90.	Leaf		√
91.	Root		√
92.	Seed germination		√
93.	Seed to seedling		√
94.	Anatomy of root		√
95.	Food web		√
96.	Ligaments		√
97.	Photosynthesis trapping light energy		√
98.	Adaptation of leaves for photosynthesis		√
99.	Prokaryota and eukaryota		√
100.	Respiratory System		√
101.	Site of photosynthesis		√
102.	Action potential		√
103.	ATP synthesis in mitochondria		√
104.	Blood Clotting		√
105.	DNA Replication		√
106.	Neuroglial cell		√
107.	Synapse		√
108.	Synaptic transmission		√
109.	The Ear (Anatomy)		√
110.	Physiology of cell membrane		√

CONTENT STANDARD: 9 - 12

Topic Name		Duration	Interactive
111.	Reproduction in fungi		√
112.	Transport across membrane		√
113.	Transport across membrane		√
114.	Viral Replication		√
115.	Cofactors		√

Earth And Space Science

1.	The Sun (Part-1)	00.05.15	
2.	The Sun (Part-2)	00.05.25	
3.	Thermal Radiation	00.03.15	
4.	Nuclear Chain Reaction	00.06.37	
5.	Gravitational Force		√
6.	Kepler's First Law of Planetary Motion		√
7.	Kepler's Third Law of Planetary Motion		√
8.	Drag Force and Terminal Velocity		√

Science in Personal and Social Perspective

1.	Drug resistance	00.03.44	√
2.	Ponds and lakes	00.02.07	√
3.	Ozone layer	00.03.37	√
4.	Bone disorder (Osteoporosis)	00.03.06	√
5.	Structure of antibody	00.03.02	√
6.	Cancer therapy (Nanotechnology)	00.04.54	√
7.	Cholesterol	00.04.02	

CONTENT STANDARD: 9 - 12

	Topic Name	Duration	Interactive
8.	Types of vaccines	00.02.11	
9.	Cancer	00.04.02	
10.	Caffeine addiction and effects	00.03.49	
11.	High calorie diet and obesity	00.03.09	
12.	Plastic recycling	00.02.31	
13.	Damage caused by UV radiation	00.01.37	
14.	Air pollution	00.02.57	
15.	Carbon cycle		√
16.	Water cycle		√
17.	Malaria		√
18.	Structure of Antibody		√

High School-Algebra

	Topic Name	Duration	Interactive
1.	$(a + b)^3$	00.02.47	√
2.	Solving an equation in one variable (Balancing method)	00.04.08	√
3.	Monomial, Binomial and Trinomial	00.03.02	
4.	Quadratic Equation: Completing the Square $(a - b)^2$	00.04.57	
5.	Quadratic Equation: Completing the Square $(a + b)^2$	00.04.22	
6.	Solution of Quadratic Equation - Illustration	00.06.43	
7.	Like and unlike terms		√
8.	Algebraic Identity : $a^2 - b^2$		√
9.	Verification : $(a + b)^2$		√

CONTENT STANDARD: 9 - 12

	Topic Name	Duration	Interactive
10.	Verification : $(a + b + c)^2$		√
	High School-Functions		
1.	Arithmetic Progression	00.05.16	√
2.	Sum of 'n' terms of an Arithmetic Progression	00.05.36	√
3.	Introduction to Sets	00.05.53	
4.	Types of Sets - I	00.05.11	
5.	Types of Sets - II	00.04.45	
6.	Function	00.05.32	
7.	General term of an Arithmetic Progression	00.03.41	
8.	Exponents and Logarithms	00.04.09	
9.	Venn Diagram		√
10.	Arithmetic Operations		√
11.	Radian		√
	High School-Geometry		
1.	Geometrical Shapes	00.04.56	√
2.	Angle	00.03.47	√
3.	Construction: Perpendicular to a line	00.03.49	√
4.	Angles in a Polygon	00.03.53	√
5.	Exterior Angles of a polygon	00.03.56	√
6.	Types of Angles formed by a Transversal	00.04.15	√
7.	Constructions: Line segment, Circle & Perpendicular Bisector	00.03.09	√

CONTENT STANDARD: 9 - 12

	Topic Name	Duration	Interactive
8.	Pythagoras' theorem	00.02.39	√
9.	Similar Triangles	00.04.06	√
10.	Cyclic Quadrilateral and its properties	00.04.50	√
11.	Area and The Perimeter of a Sector	00.04.00	√
12.	Distance Formula in 3-D	00.04.49	√
13.	Area of a Circle	00.02.39	√
14.	Volume of a Cylinder	00.04.25	√
15.	Volume of a Pyramid	00.03.50	√
16.	Volume of a Cone	00.02.56	√
17.	Volume of Frustum	00.03.49	√
18.	Coordinate Planes in Three dimensional space	00.02.59	√
19.	Visualizing Solid Shapes	00.03.47	√
20.	Measurement of Line segment and Angle	00.03.31	
21.	Property of a Rhombus	00.02.35	
22.	Tessellations	00.05.02	
23.	Geometrical shapes from Set squares	00.04.04	
24.	Construction of a Triangle - I	00.04.07	
25.	Construction of a Triangle - II	00.03.42	
26.	Construction of a Triangle - III	00.04.31	
27.	Construction of a Triangle - IV	00.03.10	
28.	Dividing a Line segment into a given ratio Internally	00.05.39	
29.	Dividing a Line segment into a given ratio Externally	00.04.00	

CONTENT STANDARD: 9 - 12

	Topic Name	Duration	Interactive
30.	Polar Coordinate System	00.03.41	
31.	Construction: Incircle	00.02.55	
32.	Construction: Direct Common Tangent	00.03.45	
33.	Construction: Indirect or Transverse Common Tangents	00.04.46	
34.	Conics an Overview	00.04.16	
35.	Locus	00.04.46	
36.	Application of Distance Formula in 3-D	00.03.30	
37.	Application of Mensuration - Cylinder	00.03.13	
38.	Symmetry		√
39.	Translation of Axes		√
40.	Angle Sum Property of Triangle		√
41.	Introduction of Trigonometry Ratios		√
42.	Trigonometric Ratios		√
43.	Application of Trigonometry		√
44.	Equation of Line in Space		√
45.	Distance Formula(2D)		√
46.	Cartesian Coordinate System		√

High School-Number and Quantity

	Topic Name	Duration	Interactive
1.	Tally Marks	00.04.04	√
2.	Introduction to Integers	00.03.28	
3.	Irrational Numbers: Geometrical Representation	00.03.48	
4.	Reading a Bar Graph	00.02.39	

CONTENT STANDARD: 9 - 12

	Topic Name	Duration	Interactive
5.	Introduction to Graphs	00.06.00	
6.	Properties of Whole Numbers (Multiplication)		√
	High School-Statistics and Probability		
1.	Probability	00.02.57	√
2.	Introduction to Combination	00.05.06	

TOTAL TOPIC IN CONTENT STANDARD- 9 - 12 - 416 24.27.05

ViziTech's Future Plans

ViziTech USA plans to coordinate with Alexandria Colonial Tours with its talented and experienced tour guides for the development of additional 3D US history lessons to go with the already 1,200 science lessons.

Making History Come Alive through 3D

Along with the 3D US history lessons from the DC area, south of DC are Jamestown, Williamsburg, and Yorktown. Alexandria Colonial Tours can coordinate the taking of videos of these historical sites to create 3D US history lessons.

North of DC is Philadelphia, Pennsylvania, which is only three hours from Washington DC. Alexandria Colonial Tours can also coordinate the taking of videos of the following sites to further expand ViziTech's 3D US history lessons:

1. National Constitution Center 17 Christ Church
2. The President's House Site 18 Betsy Ross House
3. Declaration House 19 Arch Street Friends Meeting House
4. Signers' Walk 20 Christ Church Burial Ground
5. The Liberty Bell
6. Independence Hall
7. Congress Hall
8. Old City Hall
9. Signer's Garden
10. Philosophical Hall
11. Library Hall

12. Second Bank of the United States
13. Carpenters' Hall
14. New Hall Military Museum
15. The First Bank of the United States
16. Franklin Court and B. Free Franklin Post Office

These are the potential sites in the New York City area:

1. The 911 Memorial
2. The Statue of Liberty

Funding

EdTech grants are provided directly to the state Department of Education on the basis of the state's share of funding under Part A of Title I. This is from the US Department of Education. (For other funding sources, contact wewatts2@gmail.com.)

Note: Education 3D lessons are subject to change. Some could be dropped while new ones are added.

PART II

Rethinking Business

3

MILLENNIALS AND CELL PHONE AD CAMPAIGNS

> "Reality is not the way we wish things to be nor the way things appear to be. Realty is the way they actually are. You either acknowledge reality and use it to your benefit, or it will automatically work against you."
>
> —Robert Ringer

> It's no longer the biggest organizations that will win and own and control the future. It's the fastest.
>
> —Rupert Murdoch

Technology is changing quickly, and the rapid changes are impacting businesses, schools, and churches.

Jim Carroll was named as one of the seventy-five experts in *Consumer Goods Technology* magazine "The 2020 Imperative," March 17, 2014.

Each year, *Consumer Goods Technology* magazine puts together an issue that peers into the future. Jim Carroll was named one of their esteemed visionaries in the past, and again this year for their 2020 Imperative issue.

At the AmeriQuest Symposium in Florida, Jim Carroll told the invited audience what all of us already know and feel: change is happening faster than ever before:

> Half of the global generation is under 25. They are coming into industry "wired, connected, and change oriented."
>
> The success of your organization will have nothing to do with legacy, history or size but will be defined by its ability to change. Fast.
>
> Stop clinging to that which is familiar. Begin to thrive on innovation. Think big in terms of the scope of opportunities. Start small and get familiar with the technology today. Then, finally, scale fast.
>
> Sixty-five percent of today's pre-school age children will work in jobs and careers that don't yet exist.
>
> Digital camera manufacturers have 3-6 months to sell their "new" products before they become obsolete (click!).
>
> Remember, Moore's law explains that roughly, the processing power of a computer chip doubles every 18 months while its cost cuts in half. It provides for the pretty extreme exponential growth curve we see with a lot of consumer and computer technology today.
>
> In a nutshell, my perspective on the future of education is that we need to rethink the context of "how we teach" in light of the realities that:
>
> knowledge is growing exponentially'
> the foundation of knowledge generation has forever changed
> the velocity of knowledge is accelerating
> exponential growth of knowledge leads to massive career specialization

we are in the midst of a fundamental structural organizational and career change
By 2020 or sooner, it will be all about "just-in-time knowledge."

What is this leading to?

It's leading to rapid knowledge emergence—a fundamental transformation in the role of educational institutions

In other words: much of the education structure that we have in place today doesn't match the reality of what we really need to do, given the rapid change occurring in the fundamentals of knowledge.

Younger generations are globally wired, entrepreneurial, collaborative, and change oriented. They are also now driving rapid business model change and industry transformation as they move into managerial and executive positions.

—Jim Carroll, Internet

Excerpts from Barkely's "American Millennials: Deciphering the Enigma Generation"

In 2011, the Barkley Agency partnered with The Boston Consulting Group and the Service Management Group for a comprehensive look at the entire Millennials generation. Primary Authors – Jeff Fromm, Celeste Lindell, and Lainie Decker

As defined for this study, Millennials embody the generation born between 1977 and 1995. There are 80 million of them, which makes their generation larger than the Baby Boomers (born 1946 to 1964) and three times the size of Generation X (1965–1976). They make up roughly 25% of the U.S. population.

Millennials have an annual direct spending power estimated at $200 billion. Their indirect spending power each year is approximately $500 billion, largely because of their strong influence on their parents.

Millennials include some of the earliest "digital natives." How can you best engage these early adopters of new technologies and emerging social tools?

Millennials are interested in participating in your marketing. Has your brand built a listening and participation strategy that will help you connect with your brand advocates?

Millennials are known as content creators and users. Have you enabled their creation needs in new product, marketing and customer experience design?

Millennials crave adventure—often "safer" adventures. Can you design a sense of adventure into your brand experience?

Millennials strive for a health lifestyle. Have you looked at how you balance taste with nutrition or exercise with entertainment?

Millennials seek peer affirmation. How, when and where can you engage their peers?

Millennials are "hooked" on social media in much the same was as the older generations are "hooked" on email at work. Does your brand enhance or detract from their social media experience?

Millennials are not a homogeneous cohort. Who within this group is your most influential core target and what is their mindset?

Millennials believe in cause marketing. Is your brand authentic and transparent or just using a cause to sell them something in a disingenuous way?

THE NEW PARADIGM

ENGAGEMENT—INTERACTION—CO-CREATORS

Our research shows that Millennials are 2.5 times more likely to be an early adopter of technology than older generations. Fifty-six percent of Millennials report that they are usually either one of the very first to try new technologies or are among the first group to try a new technology.

For Millennials, being an early technology adopter is not tied to life stage. Even Millennnials with children continue to adopt new technology with enthusiasm.

We believe this is a fundamental difference between generations. "Something new every six months" doesn't faze Millennials in the least—it's just how things are. They don't worry about what features will be available six months from now because there will be an even more powerful device available 12 months from now when they can afford to upgrade.

77% use a laptop at home
73% own MP3 players
67% own gaming platforms

59% have smartphones
15% use tablet computers

Social networking is important. Millennials have between 200-500 friends on Face Book. Social media isn't something they do now and then; it's an integral part of their lives and the primary way in which they communicates with their friends when they're not face to face. When they're offline too long, they feel they're missing out. Millennials are heavily influenced by their peers.

Millennials shop collaboratively. They rely more on input from social circles in making product decisions.

Millennials have a global view of travel; enjoy everything from opera to rock climbing.

To the Millennials, success and status matter. Success is a matter of hard work, and status is worth the price.

Nearly half of Millennials (45%) will go out of their way to shop at stores offering rewards programs. Forty-three percent continue to purchase the brands they grew up with, but 56% are willing to switch brands in exchange for a cents-off coupon, and 63% have purchased non-favorite brands to take advantage of a sale or promotion. Thirty-seven percent say they are willing to purchase a product or service to support a cause they believe in, even if it means paying a bit more.

With so many grocery shopping options available, it's perhaps not surprising that Millennials are moving away from the traditional grocery chains their parents still favor and opting instead for specialty, mass retailer, club and convenience stores.

Forty-two percent of Millennials shop once per week, but twice as many Millennials shop more than five times a week compared to non-Millennials.

Millennials' greater shopping frequency can be explained in part by the reasons for their store visits. They tend to shop spontaneously to gather items for a recipe, satisfy a craving or buy a pre-made dish to serve. You could call it either a lack of planning or a zest for spontaneity.

Millennials walk the line between a love of cooking and being adventurous in the kitchen and an on-the-go lifestyle that often means they're eating on the run. Thirty-nine percent of them say they prefer picking up quick meals to cooking meals. However, 64% say they love to cook and enjoy being creative in the kitchen, a far greater number than the 52% of non-Millennials who do.

Among Millennial men who are the primary grocery shopper for their household, 67% consider themselves expert or creative cooks.

In keeping with Millennials' desire to try new things, they value creative menu ideas and recipes, interesting and exotic foods, as well as samples of foods to try. Child-friendliness is important to busy Millennial moms. It's interesting to see the importance of a large deli selection, which helps fulfill the desire for pre-made and quick meals.

Female Millennials want a store experience that involves sales associates who are helpful, friendly and fashionable—who wear the store's fashions and know the product line well. She's looking for a trusted advisor who will make recommendations and offer opinions about how the fashions look when she tries them on.

Millennial men value convenience, price and availability more than anything else when they're shopping for clothing. Despite this no-nonsense approach, they still shop much more frequently than non-Millennial men.

Men are even more likely than women to take fashion cues and suggestions from sales associates, and they appreciate assistance in making their selections.

Finally, regardless of their income level, Millennials shop more often than non-Millennials and spend twice as much.

Affiliation with a cause is more important to the Millennial generation that to any previous generation.

Millennials care about causes and are more likely to show a preference toward companies that support causes—even if it means paying a bit more for those companies' products. Millennials, like non-Millennials, are interested in making a difference in the world. They believe that contributing to a cause through a company's cause marketing program is easier than doing so on their own.

What do my friends think?

Millennials will confer with family and friends (including their large social media friend groups) to help them make many of their everyday decisions. Perhaps in part because it has the potential to raise others' opinions of them, 41% of Millennials participate in cause programs by supporting friends and family in causes meaningful to those people.

Social media is key.

Millennials use social media far more regularly than older generations. Unsurprisingly, social media is one of the main ways they learn about cause initiatives. The research is undeniable that this affinity for social media is not tied to life stage, but a fundamental shift in the way this group communicates, gathers information and shares that information with others.

But don't ignore other media

Digital is clearly king, but Millennials still rely on other media, including traditional and online television, Internet radio (especially public radio) and out-of-home promotion. They are busy and multitasking, so smart brands will utilize multiple media channels in promoting their cause programs.

Millennials donate via SMS

In keeping with their digital-friendly lifestyle, Millennials use their mobile phones for almost 50% of their charitable donations. They value efficiency in many areas of their lives, so it makes sense they would apply that to their participation in cause programs as well.

Men and women donate differently

Consistent with other generations, female Millennials are more aware of cause marketing in general than their male counterparts. But men are showing more heart in recent years, so don't leave them out. Female Millennials are more likely to hear about cause programs while shopping, which means brands should make sure they are promoting their causes on their packaging and in point-of-sale materials. Women are spending more time online than their male counterparts, reinforcing the need for a big digital presence for cause programs.

More than just donations

While Millennials are sometimes willing to donate money or goods like used clothing, they prefer to volunteer their time and other alternative efforts, while non-Millennials prefer to donate money directly. Millennials don't want to stop there, however. Their social nature means they want to find ways to actively engage in cause campaigns in ways that allow them to do so with their friends or family. Brands might consider sponsoring fundraising events and other social methods of activism.

Hispanic Millennials most chcaritable of all Hispanic

Millennials are more involved in causes than are non-Hispanic Millennials. They volunteer for leadership positions at a greater pace and use texting to donate at a greater rate than non-Hispanics.

The purchase decisions of Millennials are more influenced by a company's cause campaign than those of non-Millennials—choosing an appealing cause is crucial. Millennials are considerably more likely than non-Millennials to donate via SMS text and they are inclined to make purchases that support causes they believe in." —"American Millennials: Deciphering the Enigma Generation."

For more detailed information you may want to visit the Barkley's website— HYPERLINK "https://www.barkleyus.com/millennials" https://www.barkleyus.com/millennials—and click on the link on the right—"American Millennials: Deciphering the Enigma Generation."

Barkley Agency also has a book—"Marketing to Millennials: Reach the Largest and Most Influential Generation of Consumers Ever."

Optimized Cell Phone Website—The Business Connector

The purpose is to run cell phone ad campaigns for businesses to connect with the Millennials. "Half of the global generation is under age 25. They are coming into industry "wired, connected, and change oriented." Jim Carroll

Optimization specially configures a cell phone website that tech savvy Millennials with their 200-500 Face Book friends can find on 50 different apps that are social media sites and search engines because the optimized website includes automated handset detection for the more than 5,000 mobile device configurations currently tracked. It actively places your Optimized Website at those locations on the Internet for cell phone ad campaigns.

"Millennials have an annual direct spending power estimated at $200 billion. Their indirect spending power each year is approximately $500 billion, largely because of their strong influence on their parents. It's predicted that Millennials' spending power swill increase as their earning power grows." Barkley Agency

"Today there are more wireless mobile devices being used than televisions and computers combined. Globally, there are 5.3 billion mobile cell phone subscribers—that's 77 percent of the world's population. " Apollo Bravo

Alexandria Colonial Tours Optimized Cell Phone website HYPERLINK "http://alexcolonialtours.com/" http://alexcolonialtours.com/ (Online ticket sales doubled in a month by being in the right location.) Pg. 67

PART III

Rethinking Church

4

THE RESEARCH/THE MILLENNIALS

Prologue to Rethinking Church

The church is one generation from extinction.

The Developing Gen. GAP The Active Gen. Transitioning Gen. Exiting Gen.

Ages 1–18 Missing Ages 35–50 Ages 50–65 Ages 65–80+

Generation X
The MTV Gen.
Contemporary

Christian Music

Around 80% either have left the church or are leaving the church

THE NONES
THE MILLENNIALS
Ages 18–35

The "NONES" are the religiously unaffiliated.

"The number of Americans who do not identify with any religion continues to grow at a rapid pace. One-fifth of the U.S. public—and a third of adults under 30—are religiously unaffiliated today, the highest percentages ever in Pew Research Center polling.

In the last five years alone, the unaffiliated have increased from just over 15% to just under 20% of all U.S. adults nearly 33 million people who say they have no particular religious affiliation." Pew Research

Many of the Millennials retain their Christian beliefs but have become disconnected from the church.

"In fifteen years, present trends continuing, the church in America will be half of what it is today." Stephen Mansfield

Unless the trend changes, in my opinion—based on the research—smaller churches will have to consolidate with other smaller churches to remain open or they will have to close.

One Church Researcher (David Kinnaman/Barna Group)]

The Barna Group is a visionary research and resource company located in Ventura, California. Started in 1984, the firm is widely considered to be a leading research organization focused on the intersection of faith and culture.

The Barna Group's work is relied upon by media, churches and educational institutions for insight into matters of faith and contemporary society. Its public opinion research is frequently quoted in major media outlets, such as USA Today, The Wall Street Journal, Fox News, Chicago Tribune, The New York Times, Dallas Morning News, and The Los Angeles Times.

The Barna Research Group was founded in 1984 by George Barna. As a marketing research firm, it primarily served Christian ministries, non-profit organizations and various media and financial corporations. During its nearly three decades of work, Barna Group has carefully and strategically tracked the role of faith in America, developing one of the nation's most comprehensive database on spiritual indicators.

In 2009, after starting out as an intern and working 14 years for the company, David Kinnaman acquired the Barna Group under the auspices of Issachar Companies, Inc. Issachar continues to do business as Barna Group. A 17-year research veteran, Kinnaman currently serves as the president of Barna Group, and is the majority owner of Barna Group.

Since joining Barna in 1995, David has overseen studies that have polled more than 350,000 individuals. He has designed and analyzed nearly 500 projects for a variety of clients, including the American Bible Society, Columbia House, Compassion, Easter Seals, Habitat for Humanity, Integrity Media, InterVarsity, NBC-Universal, the Salvation

Army, Sony, Thomas Nelson, Prison Fellowship, World Vision, Zondervan and many others.

As a spokesperson for the firm's research, his work has been quoted in major media outlets (such as USA Today, Fox News, CNN, Washington Post, Chicago Tribune, New York Times, Los Angeles Times, Dallas Morning News, and The Wall Street Journal). He is also in demand as a speaker about spiritual trends, teenagers and twenty-somethings, and vocation and calling.

David Kinnaman is the president and majority owner of Barna Group. He is the author of the best-selling books, "You Lost Me" and "unChristian."

David and his wife, Jill, live in Ventura, California, with their three kids." —David Kinnaman, the Barna Group

The Research: The Disconnected Millennials

According to the research studies, the twenty-first-century digital generation or the millennials—those who graduated between the years 2001 and 2012—perceive the format of the twentieth-century church as no longer relevant and are either not in the church or leaving the church. There are 80 million of them or 25 percent of the population. The question is this: how does the twentieth-century church create a format to become relevant to this twenty-first-century generation while continuing to provide a meaningful ministry to its current generations?

To help educate leaders about the Digital Generation / the Millennials and their faith journeys, the Barna team has recently completed a national tour. The series of events convened nearly 10,000 leaders, pastors and parents over the past 16 months. The conclusion was that the Digital Generation / The Millennials are leaving the church at an alarming rate. —David Kinnaman / Barna Group

The term—"NONES"—rose to prominence when a Pew Research poll found that the number of Americans who are religiously unaffiliated rose to almost 20%—a nearly 5% leap in just the last five years. One common thread in every survey has been the significant number of Millennials among these "Nones." The initial Pew survey found that nearly one-in-three members of the Millennial generation (32%) has no religious affiliation. Over half of Millennials—also identified as Nomads, Prodigals, and Exiles—between the ages of 18 and 29 with a Christian background (59%) have, at some point, dropped out of going to church after having gone regularly. Additionally, more than 50% of 18–29 year olds with a Christian background say they are less active in church compared to when they were. ‒Pew Research

Before you can find a solution, you have to admit there's a problem, and I'm the trumpet, alerting people in the church world that there's a huge problem.

—Julia Duin, author of *Quitting Church*

In fifteen years, present trends continuing, the church in America will be half of what it is today.

—Stephen Mansfield

There is a ninety-page PDF report on American millennials called "American Millennials: Deciphering the Enigma Generation." prepared by the Barkley Agency, Service Management Group, and the Boston Consulting Group. The millennials consist of eighty million people, or 25 percent of the population. This presents to businesses a 200-billion-dollar consumer market. And they are gearing advertising to connect to this market.

> "The American Millennials: Deciphering the Enigma Generation" prepared by Barkley Agency, Service Management Group, and the Boston Consulting Group.
>
> There are 80 million Millennials between the ages of 18 and 29. They make up 25% of the US population. They graduated from high school between the years of 2001 and 2012.
>
> They are larger than the Baby Boomer Generation and three times as large as
>
> Generation X – the previous generation.
>
> Less than 20% of the Millennials are in church.
>
> Our research shows that Millennials are 2.5 times more likely to be an early adopter of technology than older generations. Fifty-six percent of Millennials report that they are usually either one of the very first to try new technologies or are among the first group to try a new technology. Contrast this with non-Millennials, 35% of whom usually wait a year before trying a new technology and 22% of whom admit that they wait until a technology has become

mainstream and well established before they take the leap.

We believe the fundamental difference between generations is this:

"Something new every six months doesn't faze Millennials in the least—They don't worry about what features will be available in six months from now because there will be an even more powerful device available 12 months from now when they can afford to upgrade.

77% use a laptop computer at home
72% own MP3 players
67% own gaming platforms
59% have smartphones
15% use tablet computers, compared to 6% of non-Millennials

When it comes to using the Web, Millennials are "always on." Their access to multiple Web-enabled devices, at home and on the go, makes them power users of the Internet.

7. Social Media: Millennials have 200–500 friends on Facebook with some having over 500. Most non-Millennials have 50–100 Facebook friends.

Millennials value Social Media. One Millennial said in the interview, "I feel like I'm missing something if I'm not on Facebook every day.

My life feels richer now that I am connected to more people through Social Media. Millennials are heavily influenced by their peers. They seek peer input and affirmations on their decisions.

8. Affiliation with a cause is more important to the Millennials generation than it was to previous generations.

Millennials—as a group are digital-savvy and accustomed to getting information from a variety of sources. This opens up a broad range of marketing methods and creative opportunities for organizations to engage them.

What do Millennials think?

Millennials will confer with family and friends (including their large social media friend groups) to help them make many of their everyday decisions. Perhaps in part because it has the potential to raise others' opinions of them, 41% of Millennials participate in cause programs by supporting friends and family in causes meaningful to those people.

Social media is the key. Millennials use social media far more regularly than older generations. Unsurprisingly, social media is one of the main ways they learn about cause initiatives. The research is undeniable that this affinity for social media is not tied to life stage, but a fundamental shift in the way this group communicates, gathers information and shares that information with others.

Millennials donate via SMS. In keeping with their digital-friendly lifestyle, Millennials use their mobile phones for almost 50% of their charitable donations.

They value efficiency in many areas of their lives, so it makes sense they would apply that to their participation in cause programs as well.

More than just donations—While Millennials are sometimes willing to donate money or goods like used clothing, they prefer to volunteer their time and other alternative efforts, while non-Millennials prefer to donate money directly. Millennials don't want to stop there, however. Their social nature means they want to find ways to actively engage in cause campaigns in ways that allow them to do so with their friends or family. Churches might consider sponsoring fundraising events and other social methods of activism."

The Barkley Agency "American Millennials: Deciphering the Enigma Generation."

Additional information on the millennials in the "Rethinking Business" section of the book will have information that is also applicable to the church.

Connecting with the Millenials or The New Paradigm

For two thousand years, the local church has been the center of spiritual activity—and rightly so.

Advances in cell phone technology and 3D technology have given churches another tool to promote their messages and activities.

20th Century Church Paradigm – All Activities Are Directed To Inside The church — The "Box"

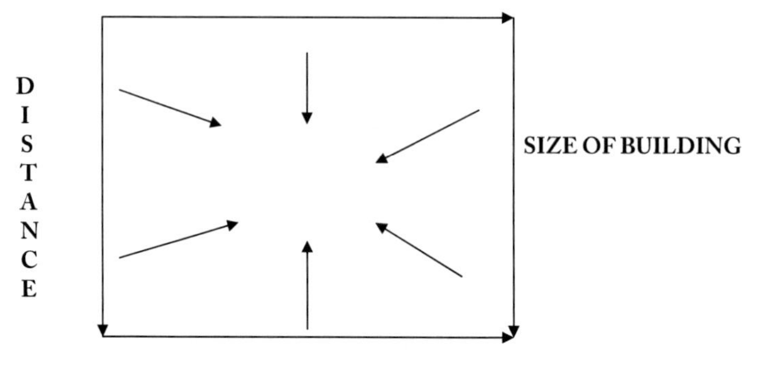

Churches can only build a finite number of buildings, and there is a limit to how far people will drive. Distance, size, and number of buildings form the twentieth-century church "box."

According to research, the millennials are trending away from the church. The challenge for the church is to connect with the millennials outside of the church.

The confluence of business technology and the church's mission meets at the cell phone. The church becomes both a center of local activity and a hub for technological outreach.

The 21st Century Paradigm – Multi-Generational – Extending Out

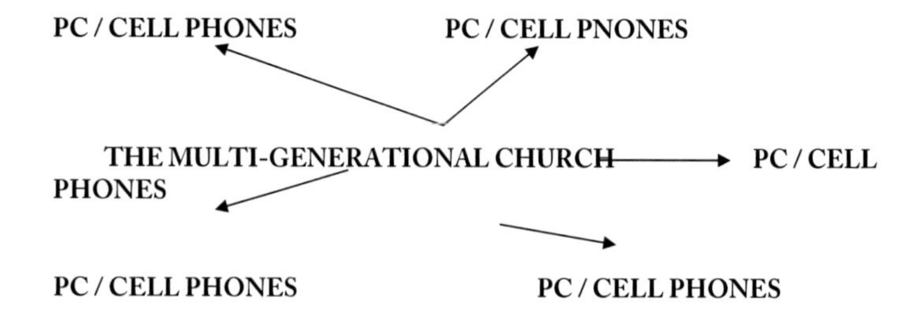

The ViziTech USA Mission

> "The Millennium generation we now deal with is considered to be a "Three Screen Generation." By that, we mean that they learn visually, from TV's, Computers, and Cell Phones, which provide multiple visual inputs, often times viewed simultaneously. We have to reach out, capture this method of input that these "digital natives" are accustomed to, and use it to reach the educational or training levels we need them to achieve."—ViziTech website

If this is true in education and business, it is also true for the church—the need "to reach out, capture this method of input that these "digital natives" are accustomed to, and use it to reach the educational or training levels we need them to achieve.

Optimized Cell Phone Website: Being in the Right Place at the Right Time

Optimization means being demographically sensitive. It means millennials can find your website on fifty different apps and websites that are social media sites and search engines. Tech companies build custom mobile sites that include automated handset detection for the more than five thousand mobile device configurations currently tracked.

The digital generation, or the millennials—a targeted demographic, ages 18–35 with their 200–500 Facebook friends—are tech savvy and use mobile devices extensively, almost exclusively.

> "Today there are more wireless mobile devices being used than televisions and computers combined. Globally, there are 5.3 billion mobile cell phone subscribers—that's 77 percent of the world's population. People in developing countries may not have televisions and computers, but most of them have cell phones."—Apollo Bravo

Here is an example of Alexandria Colonial Tours optimized cell phone website. It can also be seen on computers: HYPERLINK "http://alexcolonialtours.com/" http://alexcolonialtours.com (online ticket sales doubled in a month by being in the right location).

For more information on the optimized website, contact Wellington at HYPERLINK "mailto:wewatts2@gmail.com"wewatts2@gmail.com or Wellington at HYPERLINK "mailto:wwatts@patriot.net" wwatts@patriot.net

Diagram Showing the Utilizing of Cell Phone or Mobile Device Technology to Connect with Millennials

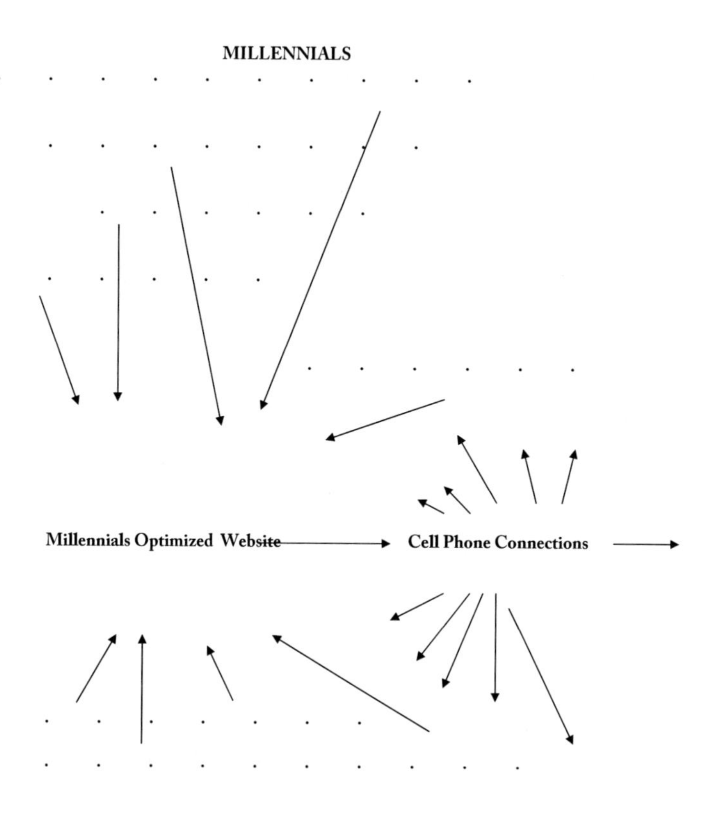

Engaging Millennials

"While Millennials are sometimes willing to donate money or goods like used clothing, they prefer to volunteer their time and other alternative efforts, while non-Millennials prefer to donate money directly. Millennials don't want to stop there, however. Their social nature means they want to find ways to actively engage in cause campaigns in ways that allow them to do so with their friends and fam-

> ily. Churches might consider sponsoring fundraising events and other social methods of activism."
> —"American Millennials: Deciphering the Enigma Generation," The Barkley Agency

One church or denomination that is doing very well in this area is the United Methodist Church. I have decided to use them as an example of the New Paradigm of Engagement.

Making Disciples of Jesus Christ for the Transformation of the World

> In the same way, faith by itself, if it is not accompanied by action, is dead.
>
> James 2:17

> The United Methodist Committee on Relief (UMCOR) is a nonprofit 501(c)3 organization dedicated to alleviating human suffering around the globe. UMCOR's work reaches people in more than 80 countries, including the United States. We provide humanitarian relief when war, conflict, or natural disaster disrupt life to such an extent that communities are unable to recover on their own.
>
> UMCOR is a ministry of The United Methodist Church, and our goal is to assist the most vulnerable persons affected by crisis or chronic need without regard to their race, religion, gender, or sexual orientation. We believe all people have God-given worth and dignity.
>
> While UMCOR cooperates with other aid organizations to extend our reach, our most important partners are the people we serve. We are con-

fident that successful solutions to emergency or chronic conditions begin with the affected population. UMCOR provides these survivors not only temporary relief but long-term education, training, and support.

UMCOR spends 100 percent of designated donations on the projects our donors specify. When UMCOR donors give their time, money, and supplies, they join UMCOR as the hands and feet of Christ.—UMCOR Website

A Future with Hope Mission Fund Campaign
Launches in GNJ

GNJ is the Greater New Jersey conference of the United Methodist Church

A Future with Hope Mission Fund Campaign has been launched with encouraging early results. This campaign, which is to raise $12 million to stop people from dying of malaria, assist survivors of Superstorm Sandy and support mission programs in our local churches, has already topped the $2 million in pledges and gifts received. Leading the effort is the commitment of our clergy—close to 75% of Greater New Jersey clergy have made a commitment totaling nearly $1 million.

The congregational phase of the campaign is designed to run simultaneously with the stewardship program churches generally conduct in the fall. District Superintendents have been coordinating webinars and meetings with GNJ staff to review manuals, walk through resources and answer questions.

The tools needed to conduct the campaign are on the GNJ website (www.gnjumc.org); however, churches around the conference have applied their creative and unique approaches to the program to make them fit each unique congregation.

The following are some examples of how GNJ churches are implementing the campaign:

Making string "mosquitoes" in Sunday School and sharing them with the congregation.

Hosting a car show in coordination with the city in order to raise money for the Mission Fund and awareness in the community of the need s of the community, region, and world.

Creating a graffiti wall so worshippers can answer the questions: "Who am I?" "Who is God?" and "Who are we together?" Placing prayer leaves by the front door with each leaf having the name of a Greater New Jersey Church, a country in Africa struggling with malaria, or a town in New Jersey hit hard by the storm. Throughout the day staff and visitors take a leaf and pray for the church, country or town then place it on the tree.

Creating a special event for immediately after worship that includes bands, food and games going through the afternoon. Taking a special offering the last week in October to recognize the one year anniversary of Superstorm Sandy and in March designated to Malaria.

Creating special Mission Fund gift envelopes to be included and interspersed monthly in the regular stewardship envelopes. Holding a church-wide dinner at a local banquet hall, including a silent auction where all the proceeds from the dinner will go towards the Mission Fund Campaign.

The Conference has dedicated three staff members to help congregations implement the campaign and guide them through the materials they need to successfully bring it to the congregation. Rebecca Nichols, Field Coordinator for Imagine No Malaria says, "My job is to make sure that our congregations have the resources they need. Yes, this campaign is about raising money, but equally important, it is to share information about urgent needs in our world. Everyone should get the opportunity to learn about these valuable programs and how they can help those who are hurting."

Bishop John Schol has clearly emphasized that this campaign is about second mile giving. Stewardship should come first. He also emphasizes that the needs are God-sized. "When people hurt, United Methodists help."

100% of the monies raised in this campaign go to the mission. Since Imagine No Malaria and A Future with Hope Inc. have been funded for their administrative overhead, all monies collected will go directly to help save lives in Africa and build homes in New Jersey.—GNJ Website

Then the King will say to those on his right, "Come, you who are blessed by my Father; take your inheritance, the kingdom prepared for you since the creation of the world. For I was hungry and you gave me something to eat, I was thirsty and you gave me something to drink, I was a stranger and you invited me in, I needed clothes and you clothed me, I was sick and you looked after me, I was in prison and you came to visit me."

Then the righteous will answer him, "Lord, when did we see you hungry and feed you, or thirsty and give you something to drink? When did we see you a stranger and invite you in, or needing clothes and clothe you? When did we see you sick or in prison

> and go to visit you?" The King will reply, "Truly I tell you, whatever you did for one of the least of these brothers and sisters of mine, you did for me."
> —Matthew 25:34–40 (NIV)

Other Outreach Programs

By providing an advanced 3D science tutorial program, churches will have the opportunity to reach out to children in their churches and in their community.

Another Outreach Program: Multiply

McLean Bible Church in McLean, Virginia, is using a program initiated by Pastor Lon Solomon called Multiply to involve his church in an effective method of sharing the gospel message with people they may encounter. They have given me permission to share this with other churches.

> WHY MULTIPLY?
>
> We want to turn Washington, DC upside down for Christ. In order to accomplish this and fulfill our mission as a church, we need to create a culture of discipleship and evangelism at McLean Bible Church, where every person is a disciple who routinely shares their faith. Imagine the impact we could have on this city if thousands of followers of Christ were inspired to share the Gospel of our Lord Jesus Christ on a daily basis.
>
> Multiply is a wake-up call for this change, focusing on Jesus' command to go and make dis-

ciples: To multiply. We launched over 200 groups the weekend of September 28 & 29, 2013, with all of our environments engaged in Multiply including small groups, community groups, sermons and classes. New groups will start again in January, and we hope you will join one then!

Our prayer is that each person at MBC Tysons will participate by joining or leading a Multiply group. Join us in this exciting initiative for God's glory.

WHAT IS MULTIPLY?

Multiply is a book by pastor and author Francis Chan to the church-at-large as a resource designed to be used in a relational environment to equip people to carry out Jesus' command—to make disciples. The goal of the Multiply material is to help you not only know the Word, but to multiply the Gospel; not only to receive, but to reproduce.

"From the beginning of Christianity, the natural overflow of being a disciple of Jesus has always been to make disciples of Jesus. "Follow me," Jesus said, "and I will make you fishers of men" (Matthew 4:19, KJV). This was a promise: Jesus would take His disciples and turn them into disciple makers. And this was a command: He called each of His disciples to go and make disciples of all nations, baptizing them and teaching them to obey Him (Matthew 28:19–20). From the start, God's design has been for every single disciple of Jesus to make disciples who make disciples who make disciples until the Gospel spread to all peoples" (Multiply, p7).

Because discipleship is so important, every aspect of the church, including small groups, community groups and Bible studies, is currently going through this material. In addition, the weekly worship services will support this initiative with complimentary teaching.

In going through the Multiply curriculum together, our prayer is that the Holy Spirit awakens the hearts of our church family and unites us in our mission to be disciples that make an impact on secular Washington DC. Being part of a Multiply group is helping us as a church body live in biblical community with the goal of becoming like Christ through obeying, serving and sharing Him.

Another Outreach: Streaming Live

Information is available at church website: HYPERLINK "http://www.mcleanbible.org" http://www.mcleanbible.org)

PART IV

Rethinking Events

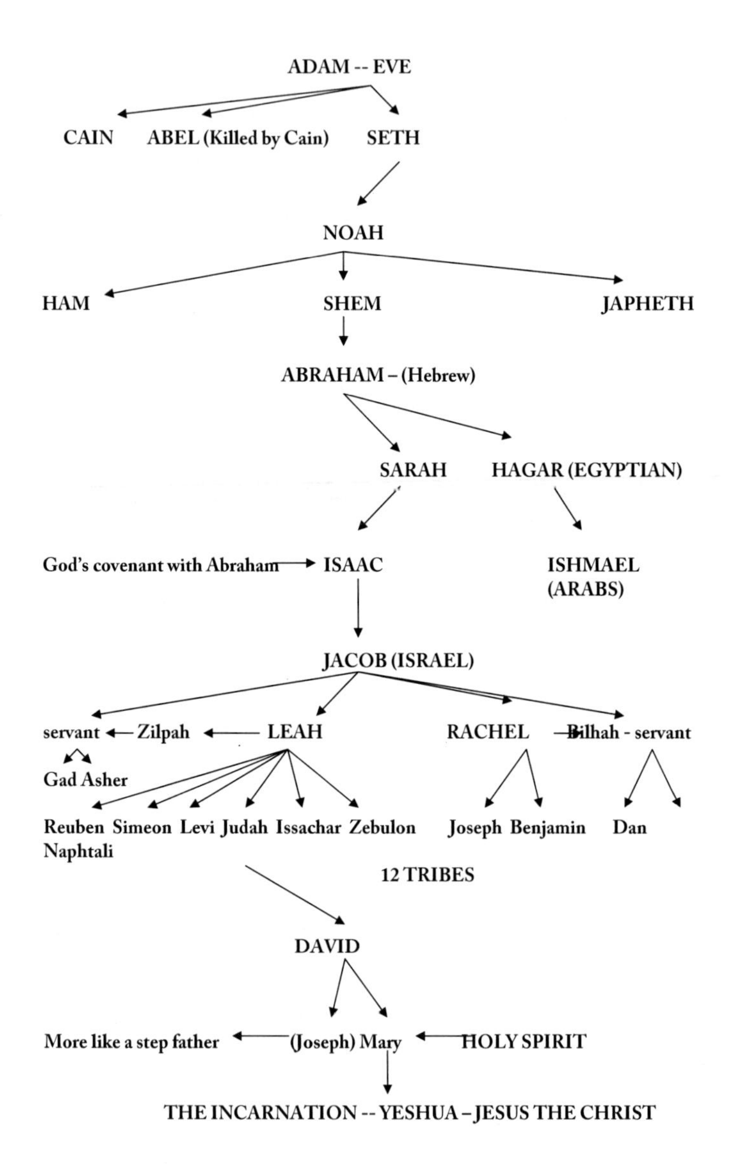
ADAM -- EVE
CAIN
ABEL (Killed by Cain)
SETH
NOAH
HAM
SHEM
JAPHETH
ABRAHAM – (Hebrew)
SARAH
HAGAR (EGYPTIAN)
God's covenant with Abraham
ISAAC
ISHMAEL
(ARABS)
JACOB (ISRAEL)
servant
Zilpah
LEAH
RACHEL
Bilhah - servant
Gad Asher
Reuben Simeon Levi Judah Issachar Zebulon
Naphtali
Joseph Benjamin
Dan
12 TRIBES
DAVID
More like a step father
(Joseph) Mary
HOLY SPIRIT
THE INCARNATION -- YESHUA – JESUS THE CHRIST

PROLOGUE TO UNDERSTANDING EVENTS AND LIFE

The Bible is the word of God. If the Bible is the word of God, then God wrote the Bible. If God wrote the Bible, then to understand the Bible, it is important to understand it from the author's—God's—point of view.

> God's plan is to repopulate a new earth with a remnant of supernaturally transformed Jews and "grafted in" Gentile believers from this earth.
>
> Romans 11

If you change your point of view, you'll see the world much differently. God is the life force. Everyone born into this world is disconnected from the life force. God's plan was to have a group of selected people to be connected to Him receiving everlasting life and to live forever on a new earth with Him. It took a couple of miracles for this to happen. God made a covenant with Abraham.

But for there to be Jewish people, God had to transform Sarah—the age of an "old grandmother"—into a "25 year old beauty queen" for Isaac to be born. From Isaac came Jacob or Israel and his twelve sons. One of his sons was Judah, and from Judah would come King David. From King David would come Mary. The second miracle was for a fourteen- or fifteen-year-old girl to have a baby without a man. That baby would be an incarnation, Yeshua—the Jewish embodiment of the life force. (Which was harder, the transformation of an "old grandmother" or the transformation of a virgin?)

His purpose was to connect the Jewish people to God, giving them everlasting life, and to have them live on a new earth with Him forever. Some of them believed in Him and were connected. Others didn't. Because of God's great love and His not wanting any to perish, the Gentiles who believed in Yeshua were "grafted in" to take the place of the Jews who didn't—but only for a time (Romans 11). But the time will come—and it may be sooner than we think—that the Jews will say, "Aha!"—we get it—and will be given everlasting life too. Then the times of the Gentiles will end, and a new order, believing Jews and grafted-in Gentiles, will begin on a new earth forever.

On the day of Pentecost, God sent His Spirit presence to earth and filled and empowered 120 Jewish followers of King Yeshua with His Holy Spirit. The Jews who had gathered in Jerusalem were simply amazed that they were able to hear what these people were saying in their own language. What happened is that 3,000 Jews believed the message of King Yeshua and instantly became supernaturally transformed—immortalized Jews.

Not all Jewish people accepted this message. Paul uses the analogy in Romans 11 of Jews who didn't accept the message as "branches that had been broken off." Because of God's love, Gentiles who believed the message were then "grafted" into the Jewish salvation in the space left by "the broken off branches."

As to the Gentles, Paul writes, "And you, though a wild olive shoot, have been grafted in among the others and now share in the nourishing sap from the olive root" (Romans 11:17, NIV).

Thus, the Gentiles that are "grafted into this Jewish salvation" become immortalized Jews.

The only people who will inhabit the new earth are immortalized Jews, physical Jews who have been immortalized, or Gentiles who have been "grafted in" to Jewish salvation—the Jewish "olive root."

It's not one's religion or church that gives people salvation. Salvation is the result of being "grafted in" to the Jewish "olive root." And from that "root," immortalization occurs, transforming the "grafted in branch" into an immortalized Jew serving King Yeshua—the last king of the Jews—forever. But the "grafting in" time isn't forever. And when it stops, the times of the Gentiles will end.

When I was about five or six, I believed without questioning that while I was asleep, Santa Claus delivered the presents that were under the tree that I saw the next morning after I had awakened. Before the age of three, my aunt Leah had taught me *The Night before Christmas*, and I knew it so well that I would appear to be reading it from a book, turning the pages at just the right time. And then "learning" and "reason" took over, and not only was Santa Claus abandoned but also the faith that I was so naïve to believe that in the first place. Reason replaced naivety. However, when I abandoned reason and reconnected with the five-year-old within me again, with the same naivet,y I discovered a kingdom that is not of this world and the supernatural God that created it.

Reconnect with the five-year-old child within you, and tell God that you trust Him completely in all areas of your life without question, and He will give you everlasting life by "grafting you into the Jewish olive root." God is the only one you can trust with this kind of all-inclusive faith because God will never betray that faith.

It takes the naivety of a child—a faith that completely accepts without question—to accept King Yeshua. Because I have done this, I have become "grafted in" to the Jewish salvation, making me an immortalized Jew through King Yeshua, the supernatural king of the Jews.

To the educated human mind, God's way to salvation and to immortality is illogical and unreasonable, maybe even a fantasy. But believe the "fantasy," and you'll live forever on a new earth. Don't trust me. Trust Him. Your brain will think thoughts you could never think on your own in a lifetime. This is called enlightenment.

> If you can believe in One you cannot see
> You are standing on the threshold of immortality.
>
> Wellington E. Watts II

The most important question for you to answer is this: Where will you be a hundred years from now?

God's purpose is to take people from this Earth to populate a new Earth and live forever on that new Earth. These people were originally the Jewish people. But God has allowed Gentiles for a time to take the place of the Jewish people that haven't accepted his plan. (Romans 12)

But this is only for a time, and the "times of the Gentiles" will come to an end.

If God is the Way, then the "ramp" to the Way is *faith*:

1. God will supernaturally transform the ones who believe in Him. That means a new spirit is within them that they didn't have when they were born.
2. God also will reside in the bodies of those who have been transformed. God's Spirit—the Holy Spirit—supernaturally enters the person's physical body, providing guidance, enlightenment, and companionship in death.
3. God will also supernaturally transfigure physical bodies. They will be forever young.

"For the Lord himself shall descend from heaven with a shout, with the voice of the archangel, and with the trump of God: and the dead in Christ shall rise first:

> Then we, which are alive and remain, shall be caught up together with them in the clouds, to meet the Lord in the air: and so shall we ever be with the Lord."
>
> —1 Thessalonians 4:16–17 (KJV)

This is all supernatural. I understand none of it. I can't really explain any of it. But you can experience all of it.

It's not hard to be one of the selected ones:

> "For God so loved the world that he gave his only begotten son that whosoever believes in him should

> not perish but have everlasting life." —John 3:16 (NIV)

Believe in him and let the supernatural begin in you. Where will you be a hundred years from now? There!

5

THE PERSPECTIVE OF TIME

Time's epicenter is in a person: Yeshua or Jesus.

Geography's epicenter is in a country: –Israel.

When the two epicenters merge together, the eternal begins.

To the "nones," millennials, atheists, agnostics, and religions of the world, "if you can believe in One you cannot see, you are standing on the threshold of immortality."

This is not about religion. It's about the supernatural transformation of the inner spirit being by God within the human body. It's about an emerging dual conscious awareness—the conscious awareness of the brain with the five senses and the conscious awareness of the transformed spirit being with the kingdom that is not of this world. It's being able to connect with God through King Yeshua—an incarnation of God into the body of a descendant of King David—making Him the last king of the Jews because He's still living, and this is all possible by the Holy Presence still active in this world. Because of God's great love, King Yeshua, the last king of the Jews, has extended His kingdom to include Gentiles—but this is only for a time.

It is being able to connect with the spirit being within you and accesses its knowledge.

The art of being still, the way of knowing, is the conscious awareness of God. "Be still and know that I am God." There is only here. And the omnipresent God is in no other place but here. And the only time with Him is now. You have never been any place but here. And you haven't lived any other time but now. Master the stillness in the here and now, for it is here that God is.

In the stillness (Psalm 46), in the secret place (Psalm 91), you feel Presence, and in His Presence, "there is fullness of joy" (Psalm 16:11).

In the stillness, the solitude of silence, a still small voice softly calls, "Follow Me."

Now my brain cannot comprehend this, and neither can yours. But if I could believe in Santa Claus, then I can believe in this. And it will take "Santa Claus faith," faith like that of a child's, for God to respond. The result is immortality, life everlasting.

Physical death is a parenthesis between your present physical life and your resurrected life. Life is an uninterrupted continuous experience by the spirit being within you. It is life everlasting.

To the educated human mind, God's way to salvation and to immortality is illogical and unreasonable, maybe even a fantasy.

But believe the "fantasy," and you'll live forever on a new earth. Don't trust me. Trust Him, and you will be surprised at the results. Your brain will think thoughts you could never think on your own in a lifetime. Welcome to the heavenly dimension. It's a nice place to be while you're still living here.

TIMES OF THE GENTILES

HEAVEN

ASCENSION OF KING YESHUA / JESUS

THE HOLY SPIRIT DESCENDS -- THE TIMES OF THE GENTILES BEGIN

HOLY SPIRIT – THE "WISHBONE" EVENT

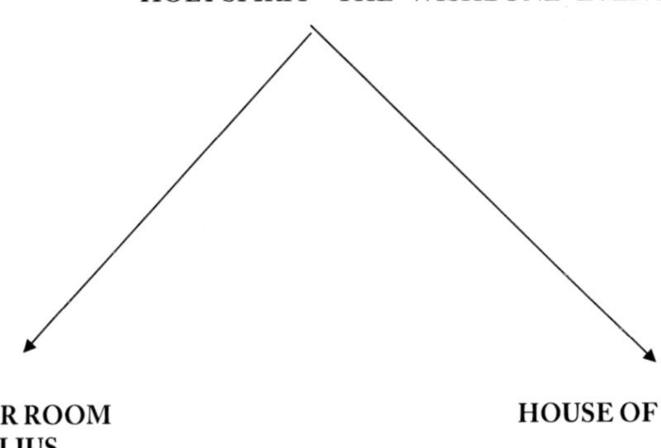

UPPER ROOM **HOUSE OF CORNELIUS**

120 DISCIPLES
3000 ADDED IN ONE DAY

TIMES OF THE GENTILES BEGIN

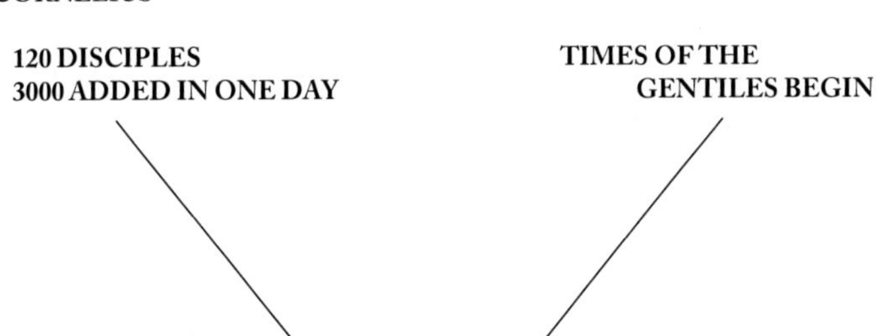

THE YESUIC KINGDOM – JEW AND GENTILE -- BEGINS

PENTECOST / TIMES OF THE GENTILES BEGIN

The Fullness of Time

The past, present, and future are impacted by history's epicenter.

This phrase carries with it the idea of a tree first blossoming. Then the fruit appears and begins to develop. Then it begins to ripen. When it is ripe, you walk up to the peach, and when you touch it, it drops off in your hand. That's the "fullness of time." The "ripening fruit" was Alexander the Great and the Roman Empire with its roads. When the "fruit was ripe," the incarnation "dropped" into this world.

"The king of Macedonia was Phillip. He had his son, Alexander, tutored by Aristotle, the Greek philosopher. Alexander soon rose to power and conquered the world of his time. It marked the triumphs of the Greek language in its Attic form. His empire lasted only for a short time. But it did mean the acceleration of the Hellenizing process, which had already begun.

When the Romans, in the last two centuries before Christ, conquered the eastern part of the Mediterranean world, they made no attempt to suppress the Greek language. On the contrary, those whom they had conquered had conquered the conquerors to a very considerable extent.

Rome herself had already come under Greek influence, and now she had made the use of the Greek language in administering at least the eastern part of her vast empire. The language of the Roman Empire was not so much Latin as it was Greek. Thus, in the first century after Christ, Greek had become a world language.

The ancient languages of the various countries did indeed continue to exist, and many districts were bilingual, the original local languages existing side by side with the Greek.

But at least in the great cities throughout the Empire, certainly in the East, the Greek language everywhere was understood. Even in Rome itself, there was a large Greek-speaking population. It is not surprising that Paul's letter to the Roman Church is written not in Latin but in Greek.

The Greek language itself underwent changes. The language of the new cosmopolitan age was very different from the original Attic dialect upon which it was founded.

This new world language, which prevailed after Alexander the Great, has been appropriately called the Koine. The word *koine* means "common"—not a bad designation, therefore, for a language, which was a common medium of exchange for diverse peoples.

The Koine is the Greek language that prevailed from about 330 BC to the close of the ancient history at about AD 500. The New Testament was written within this Koine period.

In reality, it is said the New Testament is written simply in the popular form of the Koine, which was spoken in the cities throughout the whole of the Greek-speaking world.

The New Testament writers have used the common living language of the day. But they have used it in the expression of uncommon thoughts, and the language itself, in the process, has been to some extent transformed. The cosmopolitan popular language of the Greco-Roman world served its purpose in history as well. It broke down racial and linguistic barriers. And at one point in its life became sublime.[1]

Alexander the Great provided the language. The Roman Empire provided the roads. But the fullness of time, with its "universal" language—Greek—and the Roman roads, also existed in a most brutal setting.

The Dynamics of the Times

The Culture Of Roman Decadence
The Cruelty Of Roman Punishment—Crucifixions, etc.
The Corruption Of The "Unholy" Alliance Between
Rome and the religious leaders

The fullness of time was marked by cultural decadence by the rulers—by intense cruelty to anyone who would dare to defy Rome—and by corruption between those being ruled and those ruling.

> Julius Caesar was at war. Thinking there was a truce with the Germanic tribes, Julius Caesar's cavalry was attacked in a surprise action. The Germanic attackers would jump from their horses and slit open the bellies of Caesar's cavalry's horses causing the cavalry to flee on foot—if they could. Caesar's revenge surpassed an act of war response to all out genocide killing an estimated 430,000 people. Rome is a vicious republic that gives no quarter to its enemies.[2]

> Cleopatra of Egypt had a son, which she named Caesarion. Her scheme was the Julius Caesar would become king. Needing to have a royal wife, he would marry her and adopt Caesarion as his son making him heir to the throne.

But with the assassination of Julius Caesar, everything changed. Marc Antony was now at war with Octavian, Julius Caesar's nephew. With a defection by one of his generals and his army falling apart, Marc Antony flees to Egypt knowing that he would be facing the cruelest of punishments.

The cruel punishment Marc Antony knew he would be facing if captured might have been a combination of several punishments. The usual modes of punishment were hanging, burning him alive, beheading, putting him inside a bag of scorpions and then drowning him, or crucifixion. Crucifixion was the worst and most humiliating of the punishments. Prior to the crucifixion, Marc Antony would have been stripped with his arms tied to a post. He would have been beaten with short handled whips consisting of three leather tendrils. The three tendrils were tipped with lead balls and mutton bones. These whips would lacerate the skin while tearing the skin and muscle open. The continued whipping would cause deep bruising and internal bleeding. Broken in body and in spirit, Marc Antony would have been compelled to carry the crossbeam of his cross to his place of execution.[3]

Realizing the hopelessness of his situation and not wanting to be captured and face cruel punishment, Marc Antony falls on his sword, dying in the arms of Cleopatra. Cleopatra knowing that she had no future follows Marc Antony into death by drinking a blend of poisonous opium and hemlock.

Caersarion had fled to India. But being lured back to Rome, he was promptly strangled to death. The new Roman Empire is ruled by just one all-powerful man who believes himself to be a son of a god—Octavian.

Octavian will soon answer to a new name—Caesar Augustus.[4]

And it came to pass in those days, that there went out a decree from Caesar Augustus that all the world should be taxed. (And this taxing was first made when Cyrenius was governor of Syria.)

And all went to be taxed, every one into his own city. And Joseph also went up from Galilee, out of the city of Nazareth, into Judaea, unto the city of David, which is called Bethlehem; (because he was of the house and lineage of David) to be taxed with Mary his espoused wife, being great with child.

And so it was, that, while they were there, the days were accomplished that she should be delivered. And she brought forth her firstborn son, and wrapped him in swaddling clothes, and laid him in a manger; because there was no room for them in the inn.

And there were in the same country shepherds abiding in the field, keeping watch over their flock by night. And, lo, the angel of the Lord came upon them, and the glory of the Lord shone round about them: and they were sore afraid.

And the angel said unto them, Fear not: for, behold, I bring you good tidings of great joy, which shall be to all people. For unto you is born this day in the city of David a Saviour, which is Christ the Lord.

And this shall be a sign unto you; Ye shall find the babe wrapped in swaddling clothes, lying in a manger. And suddenly there was with the angel a multitude of the heavenly host praising God, and saying, Glory to God in the highest, and on earth peace, good will toward men.

> And it came to pass, as the angels were gone away from them into heaven, the shepherds said one to another, Let us now go even unto Bethlehem, and see this thing which is come to pass, which the Lord hath made known unto us. And they came with haste, and found Mary, and Joseph, and the babe lying in a manger.—Luke 2:1–16 (KJV)

King Yeshua—an incarnation in a Jewish body, a descendant of King David—is the last of the Jewish kings because He still lives. He is currently ruling His kingdom from heaven. The Yeshuic kingdom is the only kingdom that will survive.

The birth of King Yeshua ushered in the Yeshuic kingdom into this world, which would be in direct opposition to the kingdoms and governments of this world. He also would be a potential threat to Caesar, and even a suggestion of that kind of opposition would result in death.

> After Jesus was born in Bethlehem in Judea, during the time of King Herod, Magi from the east came to Jerusalem and asked, "Where is the one who has been born king of the Jews? We saw his star when it rose and have come to worship him."
>
> When King Herod heard this he was disturbed, and all Jerusalem with him.[4] When he had called together all the people's chief priests and teachers of the law, he asked them where the Messiah was to be born. "In Bethlehem in Judea," they replied, "for this is what the prophet has written:
>
> "But you, Bethlehem, in the land of Judah,
> are by no means least among the rulers of Judah;
> for out of you will come a ruler
> who will shepherd my people Israel."

Then Herod called the Magi secretly and found out from them the exact time the star had appeared. He sent them to Bethlehem and said, "Go and search carefully for the child. As soon as you find him, report to me, so that I too may go and worship him."

After they had heard the king, they went on their way, and the star they had seen when it rose went ahead of them until it stopped over the place where the child was. When they saw the star, they were overjoyed. On coming to the house, they saw the child with his mother Mary, and they bowed down and worshiped him. Then they opened their treasures and presented him with gifts of gold, frankincense and myrrh. And having been warned in a dream not to go back to Herod, they returned to their country by another route.—Matthew 2:1–9 (NIV)

When the Magi asked Herod, "Where is He that is born King of the Jews for we have seen His star in the East and have come to worship Him?" That question brought with it a death sentence because no potential challenge to Roman authority would be allowed to live.

When Herod realized that he had been outwitted by the Magi, he was furious, and he gave orders to kill all the boys in Bethlehem and its vicinity who were two years old and under, in accordance with the time he had learned from the Magi. Then what was said through the prophet Jeremiah was fulfilled:

"A voice is heard in Ramah,
weeping and great mourning,
Rachel weeping for her children
and refusing to be comforted,
because they are no more."—Matthew 2:16–18 (NIV)

> But Mary, Joseph, and Yeshua were in Egypt having been forewarned in a dream by God.

However, the dynamics of the times would ultimately lead to the crucifixion and resurrection of King Yeshua. His ascension into heaven and the day of Pentecost marked the advent of a new era: the times of the Gentiles.

King Yeshua—the last king of the Jews, a physical descendant of King David—initiated a change. He not only is the king of the Jews but also instituted the inclusion of the Gentiles as part of the new kingdom. But the inclusion of the Gentiles is only for a time.

His death and resurrection brought reconciliation of Jew and Gentile with God.

> Who hath believed our report? and to whom is the arm of the Lord revealed?
>
> For he shall grow up before him as a tender plant, and as a root out of a dry ground: he hath no form nor comeliness; and when we shall see him, there is no beauty that we should desire him.
>
> He is despised and rejected of men; a man of sorrows, and acquainted with grief: and we hid as it were our faces from him; he was despised, and we esteemed him not.
>
> Surely he hath borne our griefs, and carried our sorrows: yet we did esteem him stricken, smitten of God, and afflicted.
>
> But he was wounded for our transgressions, he was bruised for our iniquities: the chastisement of our peace was upon him; and with his stripes we are healed.
>
> All we like sheep have gone astray; we have turned every one to his own way; and the Lord hath laid on him the iniquity of us all.

> He was oppressed, and he was afflicted, yet he opened not his mouth: he is brought as a lamb to the slaughter, and as a sheep before her shearers is dumb, so he openeth not his mouth.
>
> He was taken from prison and from judgment: and who shall declare his generation? for he was cut off out of the land of the living: for the transgression of my people was he stricken.
>
> And he made his grave with the wicked, and with the rich in his death; because he had done no violence, neither was any deceit in his mouth.
>
> Yet it pleased the Lord to bruise him; he hath put him to grief: when thou shalt make his soul an offering for sin, he shall see his seed, he shall prolong his days, and the pleasure of the Lord shall prosper in his hand.
>
> He shall see of the travail of his soul, and shall be satisfied: by his knowledge shall my righteous servant justify many; for he shall bear their iniquities.
>
> Therefore will I divide him a portion with the great, and he shall divide the spoil with the strong; because he hath poured out his soul unto death: and he was numbered with the transgressors; and he bare the sin of many, and made intercession for the transgressors. —Isaiah 53:1–12 (KJV)

Two significant events happened after Yeshua or Jesus returned to heaven:

The First Event: The First Leg of the "Wishbone"—Pentecost (see chart 4).

"When the day of Pentecost came, they were all together in one place. Suddenly a sound like the blowing of a violent wind came from heaven and filled the whole house where they were sitting. They saw what seemed to be tongues of fire that separated and came to rest on each of them. All of them were filled with the Holy Spirit and began to speak in other tongues as the Spirit enabled them.

Now there were staying in Jerusalem God-fearing Jews from every nation under heaven. When they heard this sound, a crowd came together in bewilderment, because each one heard their own language being spoken. Utterly amazed, they asked: "Aren't all these who are speaking Galileans? Then how is it that each of us hears them in our native language? Parthians, Medes and Elamites; residents of Mesopotamia, Judea and Cappadocia, Pontus and Asia, Phrygia and Pamphylia, Egypt and the parts of Libya near Cyrene; visitors from Rome (both Jews and converts to Judaism); Cretans and Arabs—we hear them declaring the wonders of God in our own tongues!" Amazed and perplexed, they asked one another, "What does this mean?"

Some, however, made fun of them and said, "They have had too much wine."

Peter Addresses the Crowd

Then Peter stood up with the Eleven, raised his voice and addressed the crowd: "Fellow Jews and all of you who live in Jerusalem, let me explain this to you;

listen carefully to what I say. These people are not drunk, as you suppose. It's only nine in the morning! No, this is what was spoken by the prophet Joel:

"In the last days, God says,
I will pour out my Spirit on all people.
Your sons and daughters will prophesy,
your young men will see visions,
your old men will dream dreams.
Even on my servants, both men and women,
I will pour out my Spirit in those days,
and they will prophesy.
I will show wonders in the heavens above
and signs on the earth below,
blood and fire and billows of smoke.
The sun will be turned to darkness
and the moon to blood
before the coming of the great and glorious day of the Lord.

And everyone who calls on the name of the Lord will be saved. "Fellow Israelites, listen to this: Jesus of Nazareth was a man accredited by God to you by miracles, wonders and signs, which God did among you through him, as you yourselves know. This man was handed over to you by God's deliberate plan and foreknowledge; and you, with the help of wicked men, put him to death by nailing him to the cross. But God raised him from the dead, freeing him from the agony of death, because it was impossible for death to keep its hold on him. David said about him:

"I saw the Lord always before me.

Because he is at my right hand,
I will not be shaken.
Therefore my heart is glad and my tongue rejoices;
my body also will rest in hope,
because you will not abandon me to the realm of the dead,
you will not let your holy one see decay.
You have made known to me the paths of life;
you will fill me with joy in your presence."

"Fellow Israelites, I can tell you confidently that the patriarch David died and was buried, and his tomb is here to this day. But he was a prophet and knew that God had promised him on oath that he would place one of his descendants on his throne. Seeing what was to come, he spoke of the resurrection of the Messiah, that he was not abandoned to the realm of the dead, nor did his body see decay. God has raised this Jesus to life, and we are all witnesses of it. Exalted to the right hand of God, he has received from the Father the promised Holy Spirit and has poured out what you now see and hear. For David did not ascend to heaven, and yet he said,

"The Lord said to my Lord:
"Sit at my right hand
until I make your enemies
a footstool for your feet."

"Therefore let all Israel be assured of this: God has made this Jesus, whom you crucified, both Lord and Messiah."

When the people heard this, they were cut to the heart and said to Peter and the other apostles, "Brothers, what shall we do?"

Peter replied, "Repent and be baptized, every one of you, in the name of Jesus Christ for the forgiveness of your sins. And you will receive the gift of the Holy Spirit. The promise is for you and your children and for all who are far off—for all whom the Lord our God will call."

With many other words he warned them; and he pleaded with them, "Save yourselves from this corrupt generation." Those who accepted his message were baptized, and about three thousand were added to their number that day. They devoted themselves to the apostles' teaching and to fellowship, to the breaking of bread and to prayer.

Everyone was filled with awe at the many wonders and signs performed by the apostles. All the believers were together and had everything in common. They sold property and possessions to give to anyone who had need.

Every day they continued to meet together in the temple courts. They broke bread in their homes and ate together with glad and sincere hearts, praising God and enjoying the favor of all the people. And the Lord added to their number daily those who were being saved.—Acts 2:1–47 (NIV)

The Second Event: "Wishbone's" Second Leg Extends to the Gentiles (see chart 4).

"At Caesarea there was a man named Cornelius, a centurion in what was known as the Italian Regiment. He and all his family were devout and God-fearing; he gave generously to those in need and prayed to God regularly. One day at about three in the afternoon he had a vision. He distinctly saw an angel of God, who came to him and said, "Cornelius!"

Cornelius stared at him in fear. "What is it, Lord?" he asked.

The angel answered, "Your prayers and gifts to the poor have come up as a memorial offering before God. Now send men to Joppa to bring back a man named Simon who is called Peter. He is staying with Simon the tanner, whose house is by the sea."

When the angel who spoke to him had gone, Cornelius called two of his servants and a devout soldier who was one of his attendants. He told them everything that had happened and sent them to Joppa. About noon the following day as they were on their journey and approaching the city, Peter went up on the roof to pray. He became hungry and wanted something to eat, and while the meal was being prepared, he fell into a trance. He saw heaven opened and something like a large sheet being let down to earth by its four corners. It contained all kinds of four-footed animals, as well as reptiles and birds. Then a voice told him, "Get up, Peter. Kill and eat."

"Surely not, Lord!" Peter replied. "I have never eaten anything impure or unclean."

The voice spoke to him a second time, "Do not call anything impure that God has made clean." This happened three times, and immediately the sheet was taken back to heaven.

While Peter was wondering about the meaning of the vision, the men sent by Cornelius found out where Simon's house was and stopped at the gate.

They called out, asking if Simon who was known as Peter was staying there.

While Peter was still thinking about the vision, the Spirit said to him, "Simon, three men are looking for you. So get up and go downstairs. Do not hesitate to go with them, for I have sent them."

Peter went down and said to the men, "I'm the one you're looking for. Why have you come?"

The men replied, "We have come from Cornelius the centurion. He is a righteous and God-fearing man, who is respected by all the Jewish people. A holy angel told him to ask you to come to his house so that he could hear what you have to say." Then Peter invited the men into the house to be his guests.

Peter at Cornelius's House

The next day Peter started out with them, and some of the believers from Joppa went along. The following day he arrived in Caesarea. Cornelius was expecting them and had called together his relatives and close friends. As Peter entered the house, Cornelius met him and fell at his feet in reverence. But Peter made him get up. "Stand up," he said, "I am only a man myself."

While talking with him, Peter went inside and found a large gathering of people. He said to them: "You are well aware that it is against our law for a Jew to associate with or visit a Gentile. But God has shown me that I should not call anyone impure or unclean. So when I was sent for, I came without raising any objection. May I ask why you sent for me?"

Cornelius answered: "Three days ago I was in my house praying at this hour, at three in the afternoon. Suddenly a man in shining clothes stood before me and said, 'Cornelius, God has heard your prayer and remembered your gifts to the poor. Send to Joppa for Simon who is called Peter. He is a guest in the home of Simon the tanner, who lives by the sea.' So I sent for you immediately, and it was good of you to come. Now we are all here in the presence of God to listen to everything the Lord has commanded you to tell us."

Then Peter began to speak: "I now realize how true it is that God does not show favoritism but accepts from every nation the one who fears him and does what is right.

You know the message God sent to the people of Israel, announcing the good news of peace through Jesus Christ, who is Lord of all. You know what has happened throughout the province of Judea, beginning in Galilee after the baptism that John preached— how God anointed Jesus of Nazareth with the Holy Spirit and power, and how he went around doing good and healing all who were under the power of the devil, because God was with him. "We are witnesses of everything he did in the country of the Jews and in Jerusalem. They killed him by hanging him on a cross, but God raised him from the dead on the third day and caused him to be seen.

He was not seen by all the people, but by witnesses whom God had already chosen—by us who ate and drank with him after he rose from the dead. He commanded us to preach to the people and to testify that he is the one whom God appointed as judge of the living and the dead. All the prophets testify about him that everyone who believes in him receives forgiveness of sins through his name.

While Peter was still speaking these words, the Holy Spirit came on all who heard the message.

> The circumcised believers who had come with Peter were astonished that the gift of the Holy Spirit had been poured out even on Gentiles. For they heard them speaking in tongues and praising God. Then Peter said, "Surely no one can stand in the way of their being baptized with water. They have received the Holy Spirit just as we have." So he ordered that they be baptized in the name of Jesus Christ. Then they asked Peter to stay with them for a few days. —Acts 10:1–48 (NIV)

The Return of King Yeshua
Signs of Yeshua's Return
End of Time: The Times of the Gentiles Come to an End

> "As Jesus was sitting on the Mount of Olives, the disciples came to him privately. "Tell us," they said, "when will this happen, and what will be the sign of your coming and of the end of the age?"
>
> Jesus answered: "Watch out that no one deceives you. For many will come in my name, claiming, 'I am the Messiah,' and will deceive many. You will hear of wars and rumors of wars, but see to it that you are not alarmed. Such things must happen, but the end is still to come. Nation will rise against nation, and kingdom against kingdom. There will be famines and earthquakes in various places. All these are the beginning of birth pains.
>
> "Then you will be handed over to be persecuted and put to death, and you will be hated by all nations because of me. At that time many will

turn away from the faith and will betray and hate each other, and many false prophets will appear and deceive many people. Because of the increase of wickedness, the love of most will grow cold, but the one who stands firm to the end will be saved. And this gospel of the kingdom will be preached in the whole world as a testimony to all nations, and then the end will come.

"So when you see standing in the holy place 'the abomination that causes desolation,' spoken of through the prophet Daniel—let the reader understand—then let those who are in Judea flee to the mountains. Let no one on the housetop go down to take anything out of the house. Let no one in the field go back to get their cloak. How dreadful it will be in those days for pregnant women and nursing mothers! Pray that your flight will not take place in winter or on the Sabbath. For then there will be great distress, unequaled from the beginning of the world until now—and never to be equaled again.

"If those days had not been cut short, no one would survive, but for the sake of the elect those days will be shortened.

At that time if anyone says to you, 'Look, here is the Messiah!' or, 'There he is!' do not believe it. For false messiahs and false prophets will appear and perform great signs and wonders to deceive, if possible, even the elect. See, I have told you ahead of time.

"So if anyone tells you, 'There he is, out in the wilderness,' do not go out; or, 'Here he is, in the inner rooms,' do not believe it. For as lightning that comes from the east is visible even in the west, so will be the coming of the Son of Man. Wherever there is a carcass, there the vultures will gather.

"Immediately after the distress of those days "'the sun will be darkened, and the moon will not give its light; the stars will fall from the sky, and the heavenly bodies will be shaken."

"Then will appear the sign of the Son of Man in heaven. And then all the peoples of the earth will mourn when they see the Son of Man coming on the clouds of heaven, with power and great glory. And he will send his angels with a loud trumpet call, and they will gather his elect from the four winds, from one end of the heavens to the other.

"Now learn this lesson from the fig tree: As soon as its twigs get tender and its leaves come out, you know that summer is near. Even so, when you see all these things, you know that it is near, right at the door. Truly I tell you, this generation will certainly not pass away until all these things have happened. Heaven and earth will pass away, but my words will never pass away.

"But about that day or hour no one knows, not even the angels in heaven, nor the Son, but only the Father. As it was in the days of Noah, so it will be at the coming of the Son of Man. For in the days before the flood, people were eating and drinking, marrying and giving in marriage, up to the day Noah entered the ark; and they knew nothing about what would happen until the flood came and took them all away. That is how it will be at the coming of the Son of Man. Two men will be in the field; one will be taken and the other left. Two women will be grinding with a hand mill; one will be taken and the other left.

"Therefore keep watch, because you do not know on what day your Lord will come. But understand this: If the owner of the house had known at what

> time of night the thief was coming, he would have kept watch and would not have let his house be broken into. So you also must be ready, because the Son of Man will come at an hour when you do not expect him. —Matthew 24:3–44 (NIV)

Nearing the End?

With a new election and with new leaders emerging throughout the world, events are moving closer to the end of time. Iran is speeding up its production of nuclear capabilities. Former president Ahmadinejad is looking for the return of Jesus Christ. How do I know this? I know this because he stated it in his last speech to the United Nations. This is important because it shows how many of their leaders think.

If you watched the speech of Ahmadinejad, you heard his reference to both the return of the Twelfth Imam and Jesus Christ. Here's an excerpt from that speech:

> God Almighty has promised us a man of kindness, a man who loves people and loves absolute justice, a man who is a perfect human being and is named Imam A1-Mahdi, a man who will come in the company of Jesus Christ (PBUH) and the righteous.

Here is another excerpt from the ending of Ahmadinejad's speech:

> The arrival of the Ultimate Savior, Jesus Christ and the Righteous will bring about an eternally bright future for mankind, not by force or waging wars but through thought awakening and developing kind-

ness in everyone. Their arrival will breathe a new life in the cold and frozen body of the world. He will bless humanity with a spring that puts an end to our winter of ignorance, poverty and war with the tidings of a season of blooming.

Now we can sense the sweet scent and the soulful breeze of the spring, a spring that has just begun and doesn't belong to a specific race, ethnicity, nation or a region, a spring that will soon reach all the territories in Asia, Europe, Africa and the US.

He will be the spring of all the justice-seekers, freedom-lovers and the followers of heavenly prophets. He will be the spring of humanity and the greenery of all ages.

Let us join hands and clear the way for his eventual arrival with empathy and cooperation, in harmony and unity. Let us march on this path to salvation for the thirsty souls of humanity to taste immortal joy and grace.

Long live this spring, long live this spring and long live this spring. Thank you.

HYPERLINK

"http://www.fourwinds10.net/siterun_data/government/united_nations/news.php?q=1348577127"

http://www.fourwinds10.net/siterun_data/government/united_nations/news.php?q=1348577127

Former President Ahmadinejad was also spiritually motivated. He believed that by starting a war with Israel the result will be the return of Jesus Christ and the Twelfth Imam. The uncertain question is will the new president of Iran believe the same and continue those policies? — Ahmadinejad's UN Speech, Internet

Yeshua or Jesus Is Making Himself Known

> In the last days, God says,
> I will pour out my Spirit on all people.
> Your sons and daughters will prophesy,
> your young men will see visions,
> your old men will dream dreams.
> Even on my servants, both men and women,
> I will pour out my Spirit in those days,
> and they will prophesy.—Acts 2:17–18 (NIV)

Lon Solomon was an Orthodox Jew and a drug user and pusher at the University of North Carolina, but God miraculously transformed his life, and now he is the pastor of a large church—Mclean Bible Church in McLean, Virginia, with three additional campuses in the DC area.

If you happen to be home on a Sunday between 10:30 a.m. and 10:40 a.m., you can be part of his service and hear him preach live by clicking on this link: http://live.mcleanbible.org. You can also read his complete testimony or listen to it on the church website at HYPERLINK "http://www.mcleanbible.org/" http://www.mcleanbible.org/.[34]

As the times of the Gentiles come to an end, it appears that God is also using some unusual means to reach some in the Moslem world.

On Sunday, January 6, 2013, at Mclean Bible Church, a Moslem young lady gave her testimony as to how she accepted Yeshua or Jesus into her life. She was blocked out, so her identity was kept secret for her own protection because she could be killed for this.

As a Moslem, she believed that Jesus was a good prophet. But that was it, until the revelation came to her as Jesus appeared to her in a dream. The astounding news she gave is that thousands of Moslems are accepting Yeshua or Jesus because He is appearing to them in dreams and visions. http://www.mcleanbible.org/ (McLean Bible Church Website)

This coincides with what Joel Rosenberg reported in his book *Epicenter*.

I believe it also coincides with the prophesy of Joel stated by Peter on the Day of Pentecost:

> In the last days, God says,
> I will pour out my Spirit on all people.
> Your sons and daughters will prophesy,
> your young men will see visions,
> your old men will dream dreams.
> Even on my servants, both men and women,
> I will pour out my Spirit in those days,
> and they will prophesy.[36]

My prayer is that God will begin to open the eyes of the Jewish people too through similar dreams and visions revealing Yeshua or Jesus.

We are living in exciting times. Be more than a spectator; be a participant in these dynamic days.

Ezekiel's Prophesy from the Geographical "Epicenter" by Joel Rosenberg, Author of *Epicenter*

In Ezekiel 38:9, Israel will face overwhelming odds. The Russian-Islamic alliance will come against Israel. Nowhere in the text, however, does Ezekiel that any nation will come to Israel's side to defend her. The War of Gog and Magog will be unlike any other war in human history.

Ezekiel 38:20 indicates that "on that day, when Gog comes against the land of Israel, the Lord God says, "My fury will mount up in My anger. In My Zeal and in My blazing wrath, I declare that on that day there will surely be a great earthquake in the land of Israel. The fish of the sea, the birds of the heavens, the beasts of the fields, all the creeping things that creep on the earth, and all the men who are on the face of the earth will shake at My presence."

The Massive earthquake is only the beginning. "I will call for a sword on him on all My mountains. (Ezekiel 38:21 Every man's sword will be against his brother.

With pestilence and with blood I will enter into judgment with him. I will rain on him and his troops and on the many peoples who are with him, a torrential rain, with hailstones, fire and brimstone.

The devastation will be so immense that Ezekiel 39:12 states it will take a full seven months for Israel to bury all of the bodies of the enemies in her midst. The process would actually take much longer except that scores of bodies will be devoured by carnivorous birds and beasts that will be drawn to the battlefields like moths to a flame.[5]

No Time: The New Tomorrow The Merging of History's Epicenter, King Yeshua, with the Geographical Epicenter, Israel

Behold, I show you a mystery; We shall not all sleep, but we shall all be changed, In a moment, in the twinkling of an eye, at the last trump: for the trumpet shall sound, and the dead shall be raised incorruptible, and we shall be changed. —1 Corinthians 15:51–52 (NIV)

For the Lord himself will come down from heaven, with a loud command, with the voice of the archangel and with the trumpet call of God, and the dead in Christ will rise first. After that, we who are still alive and are left will be caught up together with them in the clouds to meet the Lord in the air. And so we will be with the Lord forever. Therefore encourage one another with these words. —1 Thessalonians 4:16–18 (NIV)

Then I saw "a new heaven and a new earth," for the first heaven and the first earth had passed away, and there was no longer any sea. I saw the Holy City, the new Jerusalem, coming down out of heaven from God, prepared as a bride beautifully dressed for her husband. And I heard a loud voice from the throne saying, "Look! God's dwelling place is now among the people, and he will dwell with them. They will be his people, and God himself will be with them and be their God. 'He will wipe every tear from their eyes. There will be no more death or mourning or crying or pain, for the old order of things has passed away."

He who was seated on the throne said, "I am making everything new!" Then he said, "Write this down, for these words are trustworthy and true."

He said to me: "It is done. I am the Alpha and the Omega, the Beginning and the End. To the thirsty I will give water without cost from the spring of the water of life. Those who are victorious will inherit all this, and I will be their God and they will be my children. But the cowardly, the unbelieving, the vile, the murderers, the sexually immoral, those who practice magic arts, the idolaters and all liars—they will be consigned to the fiery lake of burning sulfur. This is the second death."

One of the seven angels who had the seven bowls full of the seven last plagues came and said to me, "Come, I will show you the bride, the wife of the Lamb." And he carried me away in the Spirit to a mountain great and high, and showed me the Holy City, Jerusalem, coming down out of heaven from God. It shone with the glory of God, and its brilliance was like that of a very precious jewel, like a jasper, clear as crystal. It had a great, high wall with twelve gates, and with twelve angels at the gates. On the gates were written the names of the twelve tribes of Israel. There were three gates on the east, three on the north, three on the south and three on the west. The wall of the city had twelve foundations, and on them were the names of the twelve apostles of the Lamb.

The angel who talked with me had a measuring rod of gold to measure the city, its gates and its walls. The city was laid out like a square, as long as it was wide. He measured the city with the rod and found it to be 12,000 stadia in length, and as wide and high as it is long. The angel measured the wall using human measurement, and it was 144 cubits thick. The wall was made of jasper, and the city of pure gold, as pure as glass. The foundations of the city walls were decorated with every kind of precious stone. The first foundation was jasper, the second sapphire, the third agate, the fourth emerald, the fifth onyx, the sixth ruby, the seventh chrysolite, the eighth beryl, the ninth topaz, the tenth turquoise, the eleventh jacinth, and the twelfth amethyst. The twelve gates were twelve pearls, each gate made of a single pearl. The great street of the city was of gold, as pure as transparent glass.

I did not see a temple in the city, because the Lord God Almighty and the Lamb are its temple. The city does not need the sun or the moon to shine on it, for the glory of God gives it light, and the Lamb is its lamp. The nations will walk by its light, and the kings of the earth will bring their splendor into it. On no day will its gates ever be shut, for there will be no night there. The glory and honor of the nations will be brought into it. Nothing impure will ever enter it, nor will anyone who does what is shameful or deceitful, but only those whose names are written in the Lamb's book of life. —Revelation 21:1–27 (NIV)

Then the angel showed me the river of the water of life, as clear as crystal, flowing from the throne of God and of the Lamb down the middle of the great street of the city. On each side of the river stood the tree of life, bearing twelve crops of fruit, yielding its fruit every month. And the leaves of the tree are for the healing of the nations. No longer will there be any curse. The throne of God and of the Lamb will be in the city, and his servants will serve him. They will see his face, and his name will be on their foreheads. There will be no more night. They will not need the light of a lamp or the light of the sun, for the Lord God will give them light. And they will reign for ever and ever.

John and the Angel

The angel said to me, "These words are trustworthy and true. The Lord, the God who inspires the prophets, sent his angel to show his servants the things that must soon take place."

"Look, I am coming soon! Blessed is the one who keeps the words of the prophecy written in this scroll."

I, John, am the one who heard and saw these things. And when I had heard and seen them, I fell down to worship at the feet of the angel who had been showing them to me. But he said to me, "Don't do that! I am a fellow servant with you and with your fellow prophets and with all who keep the words of this scroll. Worship God!"

Then he told me, "Do not seal up the words of the prophecy of this scroll, because the time is near. Let the one who does wrong continue to do wrong; let the vile person continue to be vile; let the one who does right continue to do right; and let the holy person continue to be holy."

"Look, I am coming soon! My reward is with me, and I will give to each person according to what they have done. I am the Alpha and the Omega, the First and the Last, the Beginning and the End.

"Blessed are those who wash their robes, that they may have the right to the tree of life and may go through the gates into the city. Outside are the dogs, those who practice magic arts, the sexually immoral, the murderers, the idolaters and everyone who loves and practices falsehood.

"I, Jesus, have sent my angel to give you this testimony for the churches. I am the Root and the Offspring of David, and the bright Morning Star."

The Spirit and the bride say, "Come!" And let the one who hears say, "Come!" Let the one who is thirsty come; and let the one who wishes take the free gift of the water of life.

I warn everyone who hears the words of the prophecy of this scroll: If anyone adds anything to them, God will add to that person the plagues described in this scroll. And if anyone takes words away from this scroll of prophecy, God will take away from that person any share in the tree of life and in the Holy City, which are described in this scroll.

He who testifies to these things says, "Yes, I am coming soon."

Amen. Come, Lord Jesus.

The grace of the Lord Jesus be with God's people. Amen.. —Revelation 22:1–20 (NIV)

PART V

Rethinking Life

6

HOW SHALL WE THEN LIVE?

A. Live With Hope
B. Live By Faith
C. Live Out Love / Compassion

What About Today?
Today—wish it would last forever
It's the only day I have
Maybe it does
Maybe it's sleep that creates the illusion of time
Somehow when I awaken from my sleep
Today has disappeared and reappeared
There was no tomorrow
Only today
TODAY—I was born
I lived
I loved
I laughed
I cried
TODAY—life began

TODAY—life was lived
TODAY—life ended
In the end—all I had was today
And now that TODAY has been transformed into immortality forever
I found peace and joy—Can't wait to see you—TODAY
May love, joy, and the peace of God be with you—Today.

Live with Hope

Heaven is real.

> "Do not let your hearts be troubled. You believe in God, believe also in me. My Father's house has many rooms; if that were not so, would I have told you that I am going there to prepare a place for you. And if I go and prepare a place for you, I will come back and take you to be with me that you also may be where I am."—John 14:1–4 (NIV)

"Heaven Is for Real" is the true story of a four-year old son of a small town Nebraska pastor who experienced heaven during emergency surgery. He talked about looking down to see the doctor operating and his dad praying in the waiting room. The family didn't know what to believe but soon the evidence was clear.

In heaven, Colton met his miscarried sister whom no one ever had told him about and his great-grandfather who died 30 years before Colton was born. He shared impossible-to-know details about each. Colton went on to describe the horse that only

Jesus could ride, about how "reaaally big" God and his chair are, and how the Holy Spirit "shoots down power" from heaven to help us.

Told by the Colton's father often in Colton's own words, the disarmingly simple message is that heaven is a real place, Jesus really loves children, and to be ready…there is a coming last battle."

"Colton had seen many pictures of Jesus, but he always said there was something wrong with them. When his father called him to look a Akiene's painting of Jesus that was being shown on a CNN Larry King interview, when asked what was wrong with that picture, his reply was that "nothing was wrong." The picture coincided with what he saw in his visit to heaven. http://heavenisforreal.net/book/ (*Heaven Is for Real*) Website

Her name is Akiane Kramarik and she was born at home underwater, on July 9, 1994, in Mount Morris, Illinois, to a stay-at-home Lithuanian homemaker mother and an American father, chef and dietary manager.

Akiane—whose name means ocean in Lithuanian—and her siblings were homeschooled for the most part and they had no television and few books, so when she began telling her family about seeing visions at age four, they were fairly certain what she was experiencing was not a result of outside influences. Her parents chose to support their daughter, which probably played a part in her prolific works.

Akiane began to sketch and write poetry at age four, advanced to painting at six and writing poetry at seven. Her first completed self-portrait sold for $10,000. A large portion of the money generated from art sales is donated by Kramarik to charities. According to Akiane, her art is inspired by her visions of heaven and her personal connection with God. "I am a self-taught painter," she told Children's Digest. "God is my teacher."

Akiane explained to her family that God gave her the visions and abilities to create her artwork and poetry, which must have come as quite a shock since both her parents were atheists at the time. They later converted to Christianity on account of Kramarik's paintings and visions. More than art was happening in their home. "Simultaneous with art was a spiritual awakening," Akiane's mother, Forelli Kramarik, told Christianity Today. "It all began to happen when she started to share her dreams and visions."

Once, according to an article in New Connexion magazine, Akiane was staring off into space, with a smile on her face and a twinkle in her eyes. Asked what she was doing, she simply answered, "I was with God again, and He told me to pray continually. He showed me where He lived. I was climbing transparent stairs; underneath I saw gushing waterfalls, and as I was approaching Him, His body was pure and intense light.

"What impressed me the most was His hands—they were gigantic! I saw no bones, or veins, no skin or blood, but maps and events. Then He told me to memorize thousands upon thousands of wisdom words on a scroll that did not look like paper, but more like intense light. And, in a few seconds, I got somehow filled up. From now on I will get up early to paint. I hope one day I'd be able to paint what I was shown."

Although she was three at the time, she'll always remember God's first message to her. "He said, 'You have to do this, and I'll help you.' He said, 'Now you can help people.' I said, 'Yes, I will.' But I said it in different words in my mind. I speak through my mind to Him," she told Christianity Today.

When asked how she knows that it's God who is speaking to her she said, "Because I can hear His voice...quiet and beautiful."

Akiane was always consumed with the faces of subjects she painted, and she found that when she prayed the right vision always appeared. When she wanted to paint Jesus, however, she spent a year mulling over her model. Finally, she asked her entire family to pray with her. The next day, a giant of a man came to her door looking for work. He was a carpenter.

Akiane immediately knew this man would be her model for her painting of Jesus. Initially the carpenter agreed, but he changed his mind. "He said that he wasn't worthy to represent his Master," Akiane told Christianity Today. "He's a Christian, and he's a humble person. But I prayed that God would change his mind and that he would call back."

The carpenter—who wishes to remain anonymous—did call Akiane back, saying that God wanted him to pose for the painting, resulting in the Jesus paintings Prince of Peace and Father Forgive Them.

The painting is startling. The eyes are loving and patient, but also piercing and fierce. He is beautiful. In fact, when Colton Burpo, the little boy who says he went to heaven at age three (see articles Part One and Part Two), saw the painting, he declared it to be the only one that ever captured what Jesus looks like. There have been many paintings since that one, though Prince of Peace is probably her most famous.

> People may wonder, "Why did Jesus choose to contact Akiane?"
>
> "I have been blessed by God," she said simply. "And if I'm blessed, there is one reason and one reason only, and that is to help others. I am donating a big portion of money to charity and to combat poverty," she said. "I want to help people. I want people to find hope in my paintings and draw people's attention to God." —Natural Religious and Spiritual Mysteries / Examiner.com
>
> National Religious and Spiritual Mysteries / Examiner.com

You may visit Akiane's gallery here: HYPERLINK "https://www.akiane.com/store" https://www.akiane.com/store.

> Praise be to the God and Father of our Lord Jesus Christ—(Yeshua)! In his great mercy he has given us new birth into a living hope [emphasis mine] through the resurrection of Jesus Christ (Yeshua) from the dead.
>
> 1 Peter 1:3

The embodiment of the *living hope*—the inner presence of Yeshua—the resurrected Son of God happens through the embodiment of God's Spirit. All the promises of God are found in the embodiment of the living hope.

It is the living hope that connects with the hope of heaven beyond you—the hope of the resurrection for you that becomes fully realized on a new earth with God forever.

Hope is a reality that has not yet been actualized or fully realized. The time lag between a future realization and a present reality is bridged by hope.

Immortal

Painting by Akiane Kramarik at age seventeen. See HYPERLINK "https://www.akiane.com/store" https://www.akiane.com/store.

Today I Stand
By Wellington E. Watts II
Comfort and Hope

Today I stand on a peaceful distant shore
In a new body—immortalized—living forever—perfect.
The atmosphere radiates waves of love

The sweet smelling scent from a myriad of beautiful flowers

Flowers I had never seen before announce the nearness of the Creator.
In the cool of this day He communes with me
Love like I had never known thrills my heart and soul.
In a brief flickering moment
All the suffering and sorrow of another time
Suffering and sorrow compared to this moment
They now seem like a pinprick.
Today I stand—
The love, joy, peace and beauty of this new world are indescribable

I feel incredibly young
Today I stand—
The evening and the morning are the first day
And it is very good!

Remembering the Immortalized Family and Friends

- Rev. W. Earl Watts
- Anna G. Watts
- Harry L. Worrell
- Nellie E. Worrell
- Wilbert Mengel
- Ethel Ruch
- Walter Schaltenbrand
- Ruth Schaltenbrand
- Wayne Schaltenbrand

- Judy Mengel
- Leah Wagner
- Ivan Watts
- Ellie Watts
- Alton Watts
- Pauline Watts
- Gordon Robertson
- Dorsey Marshall Jr.
- David Larmore
- Gladys Davis
- Nellie Grose
- JoAnne Edwards
- Gabrielle E. Oberti
- Margarethe Hannelore
- Worrell Bill Irwin
- Tommy Holshouser
- Edna Holshouser
- Kenneth Masterman
- Betty Masterman
- Ray Hoffman

> Behold, I show you a mystery; We shall not all sleep, but we shall all be changed, In a moment, in the twinkling of an eye, at the last trump: for the trumpet shall sound, and the dead shall be raised incorruptible, and we shall be changed.
>
> For this corruptible must put on incorruption, and this mortal must put on immortality.
>
> So when this corruptible shall have put on incorruption, and this mortal shall have put on immortality, then shall be brought to pass the saying that is written, Death is swallowed up in victory.
>
> O death, where is thy sting? O grave, where is thy victory?

> Therefore, my beloved brethren, be steadfast, unmovable, always abounding in the work of the Lord, forasmuch as you know that your labor is not in vain in the Lord.
>
> —1 Corinthians 15:51–55, 58 (KJV)

Live by Faith

> Behold the proud, his soul is not upright in him; but the just shall live by his faith. Habakkuk 2:4 (KJV)

What is faith? I would call it the faith of a child. I call it the Santa Claus faith. It is faith that embraces totally without question. It is a special faith reserved only for God. It's the key to heaven, immortality, and life itself—embracing God totally without question, being naive enough to believe that He has the answers for everything. It becomes a way of life. I believe that it is this kind of faith that God responds to when it is exercised. I know that I can't know, and I don't have to know. I just embrace totally without question what I may never know. (I would not trust anyone else or anything else like this except God.)

> "Now faith is confidence in what we hope for and assurance about what we do not see. This is what the ancients were commended for. By faith we understand that the universe was formed at God's command, so that what is seen was not made out of what was visible.
>
> By faith Abel brought God a better offering than Cain did. By faith he was commended as righteous, when God spoke well of his offerings. And by faith Abel still speaks, even though he is dead.

By faith Enoch was taken from this life, so that he did not experience death: "He could not be found, because God had taken him away."

For before he was taken, he was commended as one who pleased God. 6 And without faith it is impossible to please God, because anyone who comes to him must believe that he exists and that he rewards those who earnestly seek him.

By faith Noah, when warned about things not yet seen, in holy fear built an ark to save his family. By his faith he condemned the world and became heir of the righteousness that is in keeping with faith.

By faith Abraham, when called to go to a place he would later receive as his inheritance, obeyed and went, even though he did not know where he was going. By faith he made his home in the promised land like a stranger in a foreign country; he lived in tents, as did Isaac and Jacob, who were heirs with him of the same promise. For he was looking forward to the city with foundations, whose architect and builder is God. And by faith even Sarah, who was past childbearing age, was enabled to bear children because she considered him faithful who had made the promise. And so from this one man, and he as good as dead, came descendants as numerous as the stars in the sky and as countless as the sand on the seashore.

All these people were still living by faith when they died. They did not receive the things promised; they only saw them and welcomed them from a distance, admitting that they were foreigners and strangers on earth. People who say such things show that they are looking for a country of their own. If they had been thinking of the country they had left, they would have had opportunity to return. Instead, they were longing for a better country—a heavenly one. Therefore God is not ashamed to be called their God, for he has prepared a city for them.

By faith Abraham, when God tested him, offered Isaac as a sacrifice. He who had embraced the promises was about to sacrifice his one and only son, even though God had said to him, "It is through Isaac that your offspring will be reckoned." Abraham reasoned that God could even raise the dead, and so in a manner of speaking he did receive Isaac back from death.

By faith Isaac blessed Jacob and Esau in regard to their future. By faith Jacob, when he was dying, blessed each of Joseph's sons, and worshiped as he leaned on the top of his staff.

By faith Joseph, when his end was near, spoke about the exodus of the Israelites from Egypt and gave instructions concerning the burial of his bones. By faith Moses' parents hid him for three months after he was born, because they saw he was no ordinary child, and they were not afraid of the king's edict.

By faith Moses, when he had grown up, refused to be known as the son of Pharaoh's daughter. He chose to be mistreated along with the people of God rather than to enjoy the fleeting pleasures of sin.

He regarded disgrace for the sake of Christ as of greater value than the treasures of Egypt, because he was looking ahead to his reward. By faith he left Egypt, not fearing the king's anger; he persevered because he saw him who is invisible. By faith he kept the Passover and the application of blood, so that the destroyer of the firstborn would not touch the firstborn of Israel.

By faith the people passed through the Red Sea as on dry land; but when the Egyptians tried to do so, they were drowned. By faith the walls of Jericho fell, after the army had marched around them for seven days. By faith the prostitute Rahab, because she welcomed the spies, was not killed with those who were disobedient.

And what more shall I say? I do not have time to tell about Gideon, Barak, Samson and Jephthah, about David and Samuel and the prophets, who through faith conquered kingdoms, administered justice, and gained what was promised; who shut the mouths of lions, quenched the fury of the flames, and escaped the edge of the sword; whose weakness was turned to strength; and who became powerful in battle and routed foreign armies. Women received back their dead, raised to life again. There were others who were tortured, refusing to be released so that they might gain an even better resurrection. Some faced jeers and flogging, and even chains and imprisonment. They were put to death by stoning; they were sawed in two; they were killed by the sword.

They went about in sheepskins and goatskins, destitute, persecuted and mistreated— the world was not worthy of them. They wandered in deserts and mountains, living in caves and in holes in the ground.

These were all commended for their faith, yet none of them received what had been promised, since God had planned something better for us so that only together with us would they be made perfect."
—Hebrews 11:1–40 (NIV)

Live Out Love or Compassion

But when he saw the multitudes, he was moved with compassion on them, because they fainted, and were scattered abroad, as sheep having no shepherd." —Matthew 9:36 (NIV)

Master, which is the great commandment in the law?

Jesus said unto him, You shall love the Lord thy God with all your heart, and with all your soul, and with all your mind.

This is the first and great commandment.

And the second is like unto it, You shall love your neighbor as yourself."

—Matthew 22:36–39 (NIV)

"A new commandment I give unto you, That ye love one another; as I have loved you, that ye also love one another.

By this shall all men know that ye are my disciples, if ye have love one to another."

—John 13:34–35 (KJV)

"If I speak in the tongues of men or of angels, but do not have love, I am only a resounding gong or a clanging cymbal. If I have the gift of prophecy and can fathom all mysteries and all knowledge, and if I have a faith that can move mountains, but do not have love, I am nothing. If I give all I possess to the poor and give over my body to hardship that I may boast, but do not have love, I gain nothing.

Love is patient, love is kind. It does not envy, it does not boast, it is not proud. It does not dishonor others, it is not self-seeking, it is not easily angered, it keeps no record of wrongs. Love does not delight in evil but rejoices with the truth. It always protects, always trusts, always hopes, always perseveres.

Love never fails. But where there are prophecies, they will cease; where there are tongues, they will be stilled; where there is knowledge, it will pass away. For we know in part and we prophesy in part, but when completeness comes, what is in part disappears.

When I was a child, I talked like a child, I thought like a child, I reasoned like a child. When I became a man, I put the ways of childhood behind me. For now we see only a reflection as in a mirror; then we shall see face to face. Now I know in part; then I shall know fully, even as I am fully known.

And now these three remain: faith, hope and love. But the greatest of these is love.

—1 Corinthians 13:1–13 (NIV)

Then the King will say to those on his right, 'Come, you who are blessed by my Father; take your inheritance, the kingdom prepared for you since the creation of the world. For I was hungry and you gave me something to eat, I was thirsty and you gave me something to drink, I was a stranger and you invited me in, I needed clothes and you clothed me, I was sick and you looked after me, I was in prison and you came to visit me.'

Then the righteous will answer him, 'Lord, when did we see you hungry and feed you, or thirsty and give you something to drink? When did we see you a stranger and invite you in, or needing clothes and clothe you? When did we see you sick or in prison and go to visit you?' "The King will reply, 'Truly I tell you, whatever you did for one of the least of these brothers and sisters of mine, you did for me."

—Matthew 25:34–40 (NIV)

"Blessed are the peacemakers: for they shall be called the children of God."

—Matthew 5:9 (KJV)

"Lord, make me an instrument of your peace,
Where there is hatred, let me sow love;
Where there is injury, pardon;
Where there is doubt, faith;
Where there is despair, hope;
Where there is darkness, light;
Where there is sadness, joy.

O Divine Master,
grant that I may not so much seek to be consoled,
as to console;
to be understood, as to understand;
to be loved, as to love.
For it is in giving that we receive.
It is in pardoning that we are pardoned,
and it is in dying that we are born to Eternal Life."

The Prayer of Saint Francis is a Catholic Christian prayer. It is attributed to the 13th-century saint Francis of Assisi, although the prayer in its present form cannot be traced back further than 1912, when it was printed in France in French, in a small spiritual magazine called La Clochette (The Little Bell) as an anonymous prayer, as demonstrated by Dr. Christian Renoux in 2001.

The prayer has been known in the United States since 1927 when its first known translation in English appeared in January 1927 in the Quaker magazine Friends' Intelligencer (Philadelphia), where it was attributed to St. Francis of Assisi. Cardinal Francis Spellman and Senator Albert W. Hawkes distributed millions of copies of the prayer during and just after World War II – Wikipedia / Internet

Living out love is living out Christ so that others may see the "Christ in you, the hope of glory..."— Colossians 1:27 (KJV)

They might express it this way:

To me, it wasn't the truth you taught
to you so real—to me so dim

Bur when you came to me
You brought a sense of Him

For from your eyes He beckoned me
And from you heart His love was shed

Until I lost al sight of you
And saw the Christ instead.

Author Unknown

And now abides faith, hope, love, these three; but the greatest of these is love.

I Corinthians 13:13

Share the Good News with this simple message: we can change the world through e-mails and through social media—Facebook—if everyone would copy this message and post it for their friends, everyone passing it on:

> Immortality
>
> Tell God you trust Him completely in all areas of your life without question, and He will give you everlasting life. Enlightenment will follow.

That's it. Welcome to the kingdom that is not of this world.

Sometimes after the dinner is over, after the lights have been dimmed, in the quiet and beauty of solitude, you can hear the soft messages from that which is eternal—of past family members who are no longer present, who once sat where you sat. It is also your future where your children and grandchildren will someday see where you sat, and may they pause for a moment and be thankful because you were there.

Serenity

By Wellington Watts

I strolled along a summer shore,
The chilly foam from breaking waves,
Soothed my aching feet,
I saw a bright, red glowing sun,
Try to hide—where sky and ocean meet.
No artist could create this scene,
A brilliant sky with shades of red, orange, purple, and blue
And then two squawking gulls reminded me,
"Go home—it's getting dark."
They were right, of course, but did they have to be so rude?
The sun dipped below the sea—and colors faded into night,
But the gulls were wrong! For soon I saw another light,
A silver glow—and even waves reflected back,
This soft fluorescent gleam.
An ocean breeze brushed past my cheek,
The moon in front to guide,
And then I heard my soul cry out—

"When I consider your heavens, the works of your fingers,
The moon and stars which you have ordained,
What is man that you are mindful of him
And the son of man that you visit him?"
How excellent are your ways, O, Lord
How excellent are your ways.
Serenity? Sharing a peaceful moment with God.

ABOUT THE AUTHOR

My name is Wellington E. Watts II, and I am from a pastor's family. I am continuing the spirit of that ministry but in the areas of Christian education, writing, and music.

I am married to Nancy (Worrell) Watts and have one son, who is currently living in Virginia. I graduated from Eastern Pilgrim College with a BS degree. I was educated to be a pastor in the Wesleyan Church. I didn't feel a calling to pastoral ministry, so I completed education courses at Glassboro State College for permanent K-8 New Jersey certification.

My thirty-two years in education have been divided between being a public school teacher, a Christian schoolteacher, and in being the principal of a Christian school—fifteen years. Being principal of Ambassador Christian Academy enabled me to combine my pastoral education and education and experience. As principal, I continued to raise the academic standards at the academy, and our students responded with high scores on standardized tests.

I have worked in churches as a music director, choir director, and a worship team leader. I have conducted and arranged many musical productions. I have also created musical dramatizations, directed special programs, and directed community choirs. The Unity Music Singers, a community choir, was formed in 1998. Unity Music, with the choir, presented the multimedia program "A Salute to Heroes: We Will Never Forget" five times in 1999.

I wrote the music and helped produce the first tape and CD project for a talented Christian artist.

I was featured in an article written by Kevin Riordan of the *Courier-Post* because of a song I wrote as a result of the September 11 tragedy. It is called, "Let the Light of Freedom Shine."

I am also a published author. My first book was *The Cosmic Connection: Beyond the Brain.* My second book is *The Impact of Technology*.

God miraculously has brought me through two near-death experiences for a purpose.

My prayer for all who read this is: May the light of His presence shine in your eyes. May the warmth of His love emanate from your heart to the heart of others, awakening in them an intense desire to have what you have.

My first encounter with the supernatural King Yeshua entered my life at age twelve.

> For God so loved the world, that he gave his only begotten Son, that whosoever believeth in him should not perish but have everlasting life. —John 3:16 (KJV)

> Jesus said unto her, I am the resurrection, and the life: he that believeth in me, though he were dead, yet shall he live: And whosoever liveth and believeth in me shall never die. —John 11:25–26 (KJV)

> Therefore if any man be in Christ, he is a new creature: old things are passed away; behold, all things are become new. —2 Corinthians 5:7 (KJV)

> Beloved, now are we the sons of God, and it doth not yet appear what we shall be: but we know that,

> when he shall appear, we shall be like him; for we shall see him as he is.
>
> —1 John 3:2 (KJV)

Like a gentle summer breeze rustling the leaves on the swaying branches of the sturdy oaks, so the supernatural breath of God breathed into my very heart and soul, and the essence of a new creation was born—now and forever. And the spirit "born" within me is connected forever to the "kingdom that is not of this world."

My second encounter with the supernatural was when I had just crossed the street, with a briefcase in one hand and a trombone in the other on my way to school. Suddenly, I hear my name being called. "Wellington!" I swiftly turned around, thinking it was my father calling me because I had forgotten something important, because there was an urgency in the tone. When I looked back at the house, I saw nobody. But I knew I had heard a very distinctive male voice call my name.

This is my third encounter with the supernatural. This is the first time I have written or spoken in a book about this event since it happened. I believe this happened during the summer of 1959. As a teenager growing up, my bedroom was on the second floor. One hot night, I couldn't sleep and was kneeling on the floor in front of the screen in the bedroom window to keep cool (we didn't have air-conditioning then). I saw what appeared to be a young woman, slightly built, coming across the parking lot that separated the church from the parsonage. She stopped under the cherry tree outside of my window and looked up at me for about ten seconds. I couldn't make out facial features. She seemed to be completely covered in clothing. Today she would look like a Middle Eastern woman. What was unusual about her was that she wasn't walking. She appeared to be gliding across the ground. I was mesmerized by what I saw, and couldn't move.

She finally glided around the back of our house. I ran downstairs to find my father. He had been outside, taking out the trash from the back of the house to the front. When I told him about this, he said that he had seen no one.

My fourth encounter with the supernatural occurred in my first year in college:

The golden spheres of light: It began as a dream. I was walking toward steps that were extended down from the door of an airplane to the ground—like you see with Air Force One—to board the airplane. I needed to climb the staircase up to the open airplane door. Before climbing the stairs, I turned around to see a young woman with intense hatred in her eyes with clenched fists hurrying toward me. As she approached me, she said, "I hate you," as she began to pound my chest with her clenched fist. I did not resist but raised both arms straight up toward heaven and said, "But I love you."

Immediately, I felt an intense wave of love enter my body. As I began to awaken from the dream, the airport incident disappeared. Instead, I began to see golden spheres of light. The golden spheres of light would reach the middle of my forehead, and then they would explode into golden rings before my closed eyes with waves of love increasing in intensity with each exploding sphere. With each expanding wave, I began to sense an ever-increasing feeling of overwhelming love, increasing in intensity with each exploding sphere until it was too powerful to be fully realized. I began to feel like I was suffocating under its intensity.

Suddenly, I realized that tears were streaming down my face. The next thing I knew, I was on the floor, on my knees, beside my bed. I asked God to stop because I couldn't take anymore. It was the most powerful expression of love I have ever experienced. It was so powerful that I felt like I was suffocating. If it didn't stop, I thought I was actually going to die. It was too powerful for me to absorb. To say God is love would be an extreme understatement.

Here is my fifth encounter with the supernatural:

Sometimes life changes with a yes-or-no answer. It was sometime in 1965 that we were having a special service in our church of which my father was the pastor. The choir was lined up, waiting for the director to arrive. For some reason he was unable to make it. My father looked at me and said, "You direct the choir." I had never directed anything in my life, and I wasn't going to walk our on the platform of a packed sanctuary and begin now.

He wouldn't accept no and persisted in basically ordering me to do what seemed to me as an unthinkable endeavor.

I hesitated, and then, for some unknown reason, I said, "Okay." It was a simple choice—yes or no—but a simple choice that day would change my life forever.

When it was time for the choir to sing, I stood up in front of them and raised my arms. They stood up, looking at me quite quizzically. The accompanist began to play the introduction to the song.

I raised my hands, and at the moment, it felt like electricity shot through my body from head to toe. Instantly, the choir began to sing, and I directed then through the song—an old song—"We Shall Shine as the Stars of the Morning"—with its climatic ending.

I instantly could direct a choir and went on to create many choir arrangements, also directing community choirs. It was a gift!

Here is my sixth encounter with the supernatural (the instant composer):

In 1986, my father had to go to the hospital for bypass surgery because of blockages. The Thursday before he went to the hospital, our pastor, Reverend Brasco, asked me to as my wife, Nancy, to sing a special song for the Sunday's communion service.

As we were driving home from practice, her voice suddenly became very distant. In my head, I could hear music and words. Since I didn't have a piano at home, when we got home, I called my sister Marilyn and sang it to her over the phone. She played it, and Nancy sang my first song for that communion service, "In Remembrance of Me."

We went to the hospital on Tuesday, and I told my dad about my first song. He said that he'd like Nancy to sing it for him, which she did. What I didn't know at the time was that the first song that I would compose would be the last song he would hear. On November 21, 1986, he died from complications from the bypass surgery.

His life ended, but the music continued. In 1989, at the request of my wife, Nancy, I created a special song for her to sing. It was in the form of a lullaby that I called "Mary's Song." Mary was a sensitive your woman and would "ponder things in her heart." She knew what the angel had told her about this special birth, a special Son with a special destiny. And yet I believe as a young mother, she also had feelings for her firstborn son. In "Mary's Song," I have her resolving her feelings as a mother and the knowledge that He wasn't hers to keep by recognizing that even though

He was a child for the ages, for this night, "Just for tonight, You're my Son."

Richard Hartline, in a Pitman studio, beautifully recorded "Mary's Song." But it didn't end with that. For a special Christmas present, through contacts though David Young ERA Realty, my son, Wellington, made a contact with the late Frank Metis in New York City for him to write the song in sheet music form for groups to sing. Frank called Nancy several times and told her that he had arranged a lot of songs for the top artists, but this was the best Christmas song he had ever heard. He said that her husband (me) was an outstanding lyric writer.

A middle-school chorus in Indiana sang his arrangement at their Christmas concert for several years. If you haven't heard of Frank Metis, this is who he is:

Frank Metis

Frank Metis studied music at New York University and has a teacher's certificate in the Schillinger System of musical composition; as a publication arranger, he has worked for many major music and book publishers. Credits include countless single-sheet music editions, more than three hundred folios, and over one thousand printed choral arrangements. He is, directly or through their publishing, affiliated and worked for many prominent pop, rock, jazz, and theater music personalities, including Paul Anka, Dave Brubeck, Judy Collins, John Denver, James Galway, Edwin Hawkins, Jimi Hendrix, Tom Leher, Kander & Ebb, Barry Manilow, George Shearing, Bruce Springsteen, Paul Simon, and John Williams. As music editor, he has worked for Stephen Sondheim (two years) on the entire vocal score of *Sweeney Todd*.

As recording arranger/producer, he had worked in the United States and in Europe on singles and albums; his recorded background music compositions are used extensively in the United States, Canada, Europe, and Japan. As author-composer his credits include various educational publications, including a contemporary series of original multiband arrangements for for pianos: "Rock Modes and Moods," Rhythm Factory," "Contemporary Choral Warm-Ups," and "Rainbows and Royalties," a creative music market course for composers and songwriters (on two audio cassettes); a designer of Rhythm Computer, a device which within the paramcter of its programmed temporal terms visually illustrates all the mathematically possible rhythm patterns (more than forty-five million) in seventeen time signatures. He formulated the multiscore combo arrangement, an orchestration playable by groups any combination of instruments. As teacher/lecturer, he conducted workshops in contemporary composition, musical creativity, and music marketing. As a songwriter, he has written more than fifty compositions (lyrics and music), some of which deal with today's social concerns. He is a member of the American Society of Composers, Authors, and Publishers (ASCAP).

His recent assignments include the following: Philip Glass, George Shearing, Bob Dylan, 10,000 Maniacs, Depeche Mode, and Tom Waits, as well as educational projects. He has also done recording arrangements for record producers in Latin America.

Frank Metis Mail Bio To Me

My seventh encounter with the supernatural or unexplained is my special dream. I cannot reveal my contacts for security reasons. You have to trust me on this one. I had a dream about an assassination attempt. I sent it to an acquaintance. I did not know at the time that this acquaintance had a contact in national security. Here is a copy of those e-mails, showing again how God moves in mysterious ways:

(Note: This was during President Bush's first term. I took me a few days to get the courage to send it to anyone, but I felt a sense of urgency to send it.)

> During this time period, I had a dream. A car was driving in a city and came to the end of a street. As it made a left turn onto another street, gunman jumped out and shot up the car. President Bush was assassinated. Vice President Cheney immediately became the next president.
>
> I woke up and immediately began to pray for the safety of President Bush wherever he was. My impression was that it wasn't in this country.
>
> I don't know why this happened, but I decided to send it to a congressman.
>
> As I thought about it, it seemed like I was a detached onlooker to the whole event as it transpired—like watching a video—but seeing the event before it actually happened.
>
> Welly

Several days later, I received a reply to this e-mail. This is that reply:

Welly,

I sent your email about the dream to a person I know.

Though I'm not suppose to even act like I know anything about anything or anyone, much less this email, I'm forwarding a part of it just so you can see how dreams are from God. And in these last days, dreams and visions are very important.

A situation, exactly like your friend dreamed, was averted during his current trip, on the very morning (Europe time) that she e-mailed it to me. I didn't open her e-mail until that afternoon, but tell her it was confirmation of what I had notified ******** keepers of, earlier at 3:30 that a.m., our time. (Also, CAUTION your friend about spilling it to Congressional types. Mostly just loose-lips, and big-hips there, but not much ability to help!)

By the time that I opened A's—— e——, it already had been averted, though I didn't get confirmation until three hours after I opened it. So, I don't know whether your friend dreamed it before it happened, during, or afterward—but their prayers doubtlessly served! It did actually get to the street/automobile level; but, because ****** had them looking for it with particulars, it was spotted and averted with arrests made—in a country which won't have much of a sense of humor about the attempt."

My eighth encounter with the supernatural or the unexplained:

My Miracle Story

I am a twenty-first–century miracle. My story begins on May 5, 2009, when I experienced what my family doctor diagnosed as an optical migraine. When this would happen, I would only see one half of a person's face. When I watched TV, I would only see half of a person's face or be able to read the words scrolling across the bottom one word at a time. However, when I told my doctor about this, he decided to have further tests done. On May 6, 2009, I had an ultrasound done of the carotid arteries.

The results indicated that I should have further testing done. On May 22, 2009, I had an MRI, MRA, and CAT scan done. The results of the tests indicated I needed to see a cardiologist.

On July 7, 2009, I saw my cardiologist. After reading the reports, he said I needed to have a nuclear stress test and echocardiogram.

On July 23, 2009, I had the tests. The results indicated I needed to have a catheterization done. On August 3, 2009, that was completed. The results indicated I needed to have a quadruple bypass surgery. The results: 100 percent blockage in a main artery—95 percent blockage at the top on both sides where the arteries loop to the left and right. There was also 85 percent blockage further down in the right artery and 75 percent blockage even further down. I would need to have quadruple bypass surgery. On August 4, 2009, at 6:00 a.m., I had this surgery. I woke up on Wednesday at 6:00 p.m. in the ICU. (I had been unconscious for thirty-six hours.)

The surgery itself was a difficult one because of the calcification of the arteries and blood vessels caused by diabetes. I was kept alive by machines for twelve hours in a very cold environment. During the surgery, I had to have blood transfusions because of the amount of bleeding. When I came from the OR to ICU, I was immediately wrapped in heated blankets to warm up my body due to the fear of hypothermia setting in because of the prolonged exposure to the cold.

In the ICU, I had poles on both sides of me with IVs going into my neck and wrists. I had a breathing tube in my throat and four tubes coming from my abdomen for draining the blood from my chest cavity. I was also catheterized for the elimination of urine. This is how my family first saw me.

I believe it was harder on them than on me because I don't remember much of this. (Again, I had been unconscious for thirty-six hours.)

The surgeon came in and talked to my wife after the surgery. He said that he did not know how blood had been getting to my heart. He said in my condition I should have been dead several years ago. He said that I had had heart disease for many years. (This had gone undetected because all of my EKGs were normal, and I had felt no symptoms.) He said that he wasn't a very religious person, but the only reason I was still alive was because of Him (meaning God), while he pointed to heaven. (I had been supernaturally kept alive according to him.) He told Nancy that I would be in significant pain and that I would probably be in ICU for eight to ten days. He told her I would be in another cardiac step-down unit for ten days and I would probably be in a cardiac rehab unit in another ten days.

I would often have visitors—doctors and nurses—from the hospital because the word around the hospital was that

I was the Miracle Man. They just wanted to stop by and see me.

I did not experience significant pain. When I was asked what my pain was on a scale of 1 to 10, I would tell them 2, and it only went up to 4 when I was being moved. At the astonishment of the surgeon, doctors, and nurses, breathing tubes, abdomen tubes, and IVs were coming out of me at a speedy rate. I was out of ICU in four days and home in eight. On the seventh day, I was walking up and down a hall and walking up and down two flights of stairs. The physical therapists said that I didn't need them anymore, and they were signing off on me. I went home the next afternoon.

For those of you who may experience challenges or know family or friends that are going through difficult times, God is still the source of miracles, and I say this from personal experience and with confidence. There is hope. By the way, instead of thirty days, I was home in eight days.

Here is my ninth encounter with the supernatural or the unexplained:

Miracle 2

On October 20, 2011, I went to my cardiologist for a checkup. His protocols for diabetics are for them to have a stress test every two years. It was supposed to be on November 3, 2011, but I was unable to keep that appointment, unknowingly increasing a serious risk by doing so.

On December 6, 2011, I had a stress test and an ultrasound of the carotids. On December 6, 2011, my cardiologist called me and said that there had been some changes in the results compared to the stress test I took after the bypass surgery in 2009. He said he felt it was necessary for me to have a catheterization. I was hesitant to do this before Christmas, but Nancy felt that I would worry about it and that it would be best to do this as soon as possible.

Therefore, on December 13, 2011, I had a conference with Dr. McCormick, a cardiovascular interventionist (stent insertion) – doctor. He said that he felt a catheterization would be necessary to determine the internal state of affairs.

On December 15, 2011, I had this done. He said the results showed that the bypasses were still good, but another artery was 90 percent blocked and would need three stents. This procedure would probably take about thirty to forty-five minutes, maybe as much as an hour.

On December 21, 2011, I had the procedure done. However, something unexpected happened. The catheterization didn't show the entire problem. At the bottom of the artery, it was 100 percent blocked.

It also showed that the heart muscle was weakening due to a lack of blood supply. Something had to be done immediately, or within a day or two I could have a heart attack and might die as a result.

The doctor assisting Dr. McCormick told Nancy that at that time he began to pray. Medically, it looked like an impossible situation. He said that a lesser doctor would have quit. It was only Dr. McCormick's dedication, determination, perseverance, and skill that continued the process for over four hours. He is one of the top doctors in the United States in this area.

The verse that came to my mind as I was on the cath table was "Thou wilt keep him in perfect peace, whose mind is stayed on thee: because he trusteth in thee" (Isaiah 26:3).

I believe it took every prayer for Dr. McCormick to be able to break through the blockage. I am so grateful for those prayers. When he finished, he said that all the stents were in place and that the blood was flowing like a river to the heart and to the vessels that had formed to get blood to the heart in compensation for the loss created by the blockage. After they had finished, the doctor told Nancy, "Here is your Christmas miracle!"

ENDNOTES

1. Introduction to New Testament Greek For Beginners by J. Gresham Machen, D.D., Litt. D – Professor of New Testament in Westminster Theological Seminary, Philadelphia – Published by the Macmillan Company – 1959
2. "Killing Jesus" – Bill O'Reilly and Martin Dugard – Henry Holt and Company New York – 2013 — pgs. 34–35
3. "Killing Jesus" – Bill O'Reilly and Martin Dugard – Henry Holt and Company New York – 2013 — pgs. 84-85
4. "Killing Jesus" – Bill O'Reilly and Martin Dugard – Henry Holt and Company New York – 2013 — pgs. 61- 63
5. "Epicenter" – Joel C. Rosenberg – Tyndale House Publishers, Inc. – Carol Stream, Illinois – 2006 – pgs 182-185

APPENDIX

VIZITECH DEMONSTRATION / RESELLER CENTER

Alexandria, Virginia

Specializing in Mobile 3D Projection Units

3D in Education

INTRODUCING THE RESELLER

My name is Wellington E. Watts III, and I am the owner of Alexandria Colonial Tours in Alexandria, Virginia. Alexandria Colonial Tours is an eighteen-year-old established business in Alexandria. I have also been officially named a Reseller of ViziTech products and have established a ViziTech office in Alexandria, Virginia.

As a reseller, I will have the nonexclusive territory of Virginia, Washington DC, and West Virginia. Non-exclusive means that no other reseller will sell in my area, and that my territory is a "starter zone," and that I don't have to sell exclusively in those areas. If I happen to have something that I can sell that is close enough to my area, then I can provide good support after the sale.

It also means that ViziTech reserves the right to sell to the US Military in my area. ViziTech is keeping that market with them, for right now, because of Stewart Rodeheaver's military ties and because he has projects going with companies headquartered in the DC area, but the projects are in Georgia, Alabama, and Colorado.

> "Once you get up to speed, we can get you involved with these types of DOD projects also."—Brigadier General Stewart Rodeheaver (retired)

As a Reseller, I sell the products of Vizitech, and I provide support to our customers. After the initial sale, I will have to teach users how to use the equipment and then check with them a couple of times a year to make sure their product is working well, their libraries are not timed out, and to try and sell them other products.

According to Stewart Rodeheaver, "It is a lot of work but the payoff is high. We sell a lot of renewals, new libraries, and new machines when we make these service calls. They become an important revenue stream."

In summary, as a reseller, I will be selling ViziTech education products, and in time I will also be handling DOD contracts out of the Alexandria office.

For a PDF Introduction to the ViziTech Reseller Center in Alexandria, Virginia, please contact Wellington at wewatts2@gmail.com